HARDWARE VERIFICATION WITH C++
A Practitioner's Handbook

HARDWARE VERIFICATION WITH C++
A Practitioner's Handbook

MIKE MINTZ
ROBERT EKENDAHL

 Springer

Cover art from the original painting "Every other now and then" by
John E. Bannon, johnebannon.com

Mike Mintz
Robert Ekendahl

Hardware Verification with C++: A Practitioner's Handbook

Library of Congress Control Number: 2006928441

ISBN 0-387-25543-5   e-ISBN 0-387-36254-1
ISBN 9780387255439

Printed on acid-free paper.

Printed in the United States of America.

9 8 7 6 5 4 3 2 1

springer.com

Contents

Contents

Contents

Part III:
Using OOP for Verification
(Best Practices). 153

Part IV:
Examples
(Putting It All Together) 271

Preface

So what makes a person write a book? Insanity is a definite probability. Another possibility is a strong desire to help. We are not special people. We have worked in several companies, large and small, made mistakes, and generally muddled through our work. There are many people in the industry who are smarter than we are, and many coworkers who are more experienced.

We have been in the lab when we bring up the chips fresh from the fab, with customers and sales breathing down our necks. We've been through software bring-up and worked on drivers that had to work around bugs, er, features, in production chips.

What we feel makes us unique is our combined broad experience from both the software and hardware worlds. Mike has over 20 years of experience from the software world that he applies in this book to hardware verification. Robert has over 12 years of experience with hardware verification, with a focus on environments and methodology.

So what we bring to the task of functional verification is over three decades of combined experience, from design, verification, software development, and management. It is our experiences that speak in this handbook. It is our desire that others might learn and benefit from these experiences.

We have had heated discussions over each line of code in this book and in our open-source libraries. We rarely agree at first, but by having to argue our cases we arrive at what we feel are smart, efficient, flexible, and simple solutions. Most of these we have "borrowed" from the software industry but have applied to the field of verification.

We believe that the verification industry can benefit from the lessons learned from the software domain. By using the industry-standard C++ language, the verification domain can adapt techniques and code from over twenty calendar years of software effort, the scope of which is nothing short of stunning. Many brilliant people have paved the way in the software field. Although the field of verification is much younger, we could benefit greatly from listening, learning, and adapting mature programming techniques to the production of products of the highest quality.

So why did we include open-source software with the handbook? Open-source software is a key to uniting and increasing the productivity of our industry.

There is almost no successful closed-source ("hard macro") intellectual property (IP), for a good reason. Without the ability to look at the source and edit as necessary, the task is much more difficult and the chances for success are slim.

You can find the open-source software and the examples used in this handbook on the accompanying CD. We've also included a simulator, which can be used for 30 days. The installer should run automatically, but in case it does not on your system, you can find the code at the top level of the CD.

The writing of this handbook has been a great and interesting, if slightly tiring, ride for us. We hope that the reading—and better yet, the application—of the basic principles inside will be of value for you.

Acknowledgments

It takes a village to raise a child, and it takes a village to create a book. There is a core family, and a few relatives, and a whole lot of helpful neighbors. The authors would like to bow humbly to our village.

This book would have been abysmal without the help of Tori Hunter and Gerry Ventura, who read through many, many early incarnations. Henrik Scheuer also provided great feedback, both detailed and "big picture." His feedback helped improve many chapters substantially.

Chris Kappler introduced Mike to hardware verification with C++ many years ago. Mike takes his hat off to Chris. David Kelly of Mentor Graphics deserves special mention. Ever since we discussed Teal with Dave several year ago, he has been a stalwart supporter.

Kate Pulnik did the graphic design. Without her our book would still have that "paper napkin scribbled on by engineers" look.

Michael Meyer was our main technical editor, turning our gibberish into English and making clear where we where unclear. This book would not have been readable without him.

We are truly grateful for all the reviewers, their time, and their suggestions during both the early and near final stages of the book. In particular we thank Ed Arthur, Per Bojsen, John DeRoo, Bob Fredieu, Ziad Hachem, Bennet Ih, Göran Knutson, Achot Matevossian, Andy Meyer, Jeff Paquette, Rex Perkins, Andrew Scougal, Brian Slater, Chris Spear, Bjarne Stroustrup, and Sterling Wight.

We can't forget Felix Unogwu and Andrew Zoneball, who really wanted to be mentioned in this book.

We are also grateful for the support and encouragement of the producers of the HDL simulators. In particular, we thank the following simulator companies—Aldec, Cadence, Mentor Graphics, and Synopsis—for providing licenses to their products, so we could confirm that the examples in this handbook work.

Introduction

Coding is a human endeavor. Forget that and all is lost.

Bjarne Stroustrup, father of C++

There are several books about hardware verification, so what makes this handbook different? Put simply, this handbook is meant to be useful in your day-to-day work. The authors are like you, cube dwellers, with battle scars from developing chips. We must cope with impossible schedules, a shortage of people to do the work, and constantly mutating hardware specifications.

We subtitled this book *A Practitioner's Handbook* because it contains real-world code examples and techniques. Sure, we talk about programming theory, but the theme of this book is writing code. We focus mainly on object-oriented programming (OOP) techniques and coding in C++. We back this up with a CD-ROM containing working, open-source Verification Intellectual Property (VIP), scripts, and several complete test systems.

We cover the following topics:

- C++ as a verification language

- The evolution of OOP, C++, and verification

- How to use OOP to build a flexible and adaptable verification system

- How to use specific OOP techniques to make verification code both simpler and more adaptable, with reference to actual situations (both good and bad) that the authors have encountered

- Useful C++ code, both as snippets, complete examples, and code libraries—all available as open source

This handbook is divided into four major sections:

- *Part I* provides an overview of OOP concepts, then walks through transforming a block-level view of a typical verification system into code and classes.

- *Part II* describes two free, open-source code libraries that can serve as a basis for a verification system—or as inspiration for your own environment. The first, called Teal, is a C++-to-HDL (hardware description language) interface. The second, called Truss, is a complete verification system framework. Both are available as open source and are included on the companion CD-ROM.

- *Part III* describes how to use OOP to make your team as productive as possible, how to communicate design intent better, and how to benefit from "lessons learned" in the software world.

- *Part IV* describes several complete real-world examples that illustrate the techniques described in the earlier parts of this book. In these examples we build complete verification environments with makefiles, scripts, and tests. These examples can serve as starting points for your own environment.

For the curious, each of the chapters in Part I and Part III ends with a section called "For Further Reading," which recommends relevant landmark papers and books from the software domain.[1]

[1.] The references in these sections, though not academically rigorous, should be sufficient to help you find the most recent versions of these works on the Internet.

Background

· ·

The silicon revolution[2] has made computers, cell phones, wireless networks, and portable MP3 players not only ubiquitous but in a constant state of evolution. However, the major impediment to introducing new hardware is no longer the hardware design phase itself, but the verification of it.

Costs of $1M or more and delays of three to six months for new hardware revisions of a large and complex application-specific integrated circuits (ASICs) are common, providing plenty of incentive to get it right the first time. Even with field-programmable gate arrays (FPGAs), upgrades are costly, and debugging an FPGA in the lab is very complex for all but the simplest designs.

For these reasons, functional verification has emerged as a team effort to ensure that a chip or system works as intended. However, functional verification means different things to different people. At the 30,000-foot level, we write specifications, make schedules, and write test plans. Mainly, though, we code. This handbook focuses on the coding part.

White papers are published almost daily to document some new verification technique. Most of you probably have several papers on your desk that you want to read. Well, now you can throw away those papers! This handbook compresses the last ten years of verification techniques into a few hundred pages. Of course, we don't actually cover that decade in detail (after all, this is not a history book), but we have picked the best techniques we found that actually worked, and reduced them to short paragraphs and examples.

Because of this compression, we cover a wide variety of topics. The handbook's sections range from talking about C++, to introducing OOP, to using OOP at a fairly sophisticated level.

---

2. Moore's law of 1965 is still largely relevant. See "Cramming more components onto integrated circuits," by Gordon Moore, *Electronics*, Volume 38, Number 8, April 19, 1965.

What is Functional Verification?

Asking "what is functional verification?" brings to mind the familiar poster, "A View of the World from Ninth Avenue,"[3] in which the streets of New York City are predominant and everything beyond is tiny and insignificant. Every one of us has a different perspective, all of which are, of course, "correct." Put simply, functional verification entails building and running software to make sure that a device under test (DUT, or in layman's terms, the chip) operates as intended—before it is mass-produced and shipped.

We perform a whole range of tasks where the end goal is to create a *high degree of confidence in the functionality of the chip*. Mostly we try to find errors of logic, by subjecting the chip to a wide variety of conditions, including error cases (where we validate graceful error handling and ensure that the chip at least does not "lock up"). We also make sure that the chip meets performance goals, and functions in uncommon combinations of parameters ("corner cases"), and confirm that the register, interrupt, and memory-map interfaces work as specified.

As with the view of New York City, the perspectives of every company, indeed even of the design and test teams within a company, will naturally be slightly different. Nevertheless, as long as the chip works as a product, there are a number of ways to achieve success. That's why this handbook does not focus on what the specific tasks are; you know what you have to do. Rather, we focus on how you can write your code as effectively as possible, to alleviate the inevitable pain of verification.

Why Focus on C++?

A major development in the field of functional verification is the increasingly mainstream use of OOP techniques. Basically, those of us in the verification field need those techniques to handle increasingly complex tasks effectively. While most of techniques presented in this handbook

---

[3.] Saul Steinberg, cover of *The New Yorker*, March 29, 1976.

are adaptable to any number of languages such as Vera or SystemVerilog, we focus on C++; the marriage of the C programming language and OOP.

At its core, OOP is designed to manage complexity. All other things being equal, simpler code is better. Because of the flexibility inherent in using OOP, we can write code that is simpler to use, and therefore more adaptable. In short, we can write reusable code that outlives its initial use.

This handbook is all about providing techniques, guidelines, and examples for using C++ in verification, allowing you to make more use of some "lessons learned" by software programmers. We distill the important bits of knowledge and techniques from the software world, and present them in the light of verification.

A Tour of the Handbook

The four parts of this handbook provide a variety of programming tips and techniques.

- *Part I* walks through the main concepts of OOP, introducing how to transform your high-level "whiteboard" idea for a verification system into separate roles and responsibilities. The goal is to build appropriately simple and adaptable verification systems.

- *Part II* uses these techniques and presents two open-source code libraries for verification called Teal and Truss. Teal is a C++-to-HDL interface (or *gasket*) that is lightweight and unobtrusive. Truss is a verification framework that encourages the use of the canonical form described in Part One. Both are used by several companies and run under most simulators.

- *Part III* introduces the OOP landscape in a fair amount of detail. OOP thinking, design, and coding are illustrated by means of code snippets representative of problems that verification engineers commonly have to solve. The chapters in Part Three provide a handbook of techniques that are used in the software world to write simple, adaptable code.

- *Part IV* provides several complete examples of verification test systems, providing real-world examples and more details on how the OOP techniques discussed are actually used. Part Four is all about code. While a handbook may not be the best vehicle for describing code, it can be a good reference tool. We start with a relatively simple example of how the verification of a single block of the ubiquitous UART[4] can be done. Then we show how this block-level environment can be expanded to a larger system.

The authors sincerely hope that, by reading this handbook, you will find useful ideas, techniques, and examples that you can use in your day-to-day verification coding efforts.

For Further Reading

- On the topic of coding well, *Writing Solid Code*, by Steve McGuire, is a good tour of the lessons Microsoft has learned.

- *Principles of Functional Verification*, by Andreas Meyer, provides an introduction to the broad topic of chip verification.

- *Writing Testbenches: Functional Verification of HDL Models, Second Edition*, by Janick Bergeron, gives another view of the process of functional verification.

[4.] Universal asynchronous receiver-transmitter.

Part I:
C++ and Verification
(The Why and How)

This part of the handbook explores the authors' choice of C++ for verification. In the next chapter we take a brief look at the other choices available, and then look at the benefits and drawbacks of using C++.

Next, we weave three different themes together: the evolution of programming in general and C++ in particular, the creation of object-oriented programming (OOP) techniques, and the evolution of functional verification. The reason we chose to look at these three themes is to show why OOP exists and how it can be harnessed to benefit verification.

A major theme of this handbook is to build a verification system in layers. C++ in general, and OOP techniques specifically, are well-suited to this approach. In the last chapter of this section, we'll look at a canonical verification system by using a standard approach to building verification components.

Why C++?

> *A language is a dialect with an army.*
> *Old proverb*

We, in the functional verification trade, write code for a living. Well, we do that, and also puzzle over code that has been written and that has yet to be written. Because functional verification is a task that only gets more complex as designs become more complex, the language we work in determines how well we can cope with this increasing complexity.

The authors believe that C++ is the most appropriate choice for functional verification, but as with any choice, there are trade-offs. This chapter discusses the advantages and disadvantages of using C++ for functional verification. We'll look at the following topics:

- An abbreviated comparison of the languages and libraries available for functional verification

- Why C++ is the best choice for verification

- The side benefits of using C++, such as the ability to share code among the software, diagnostics, and modeling teams

- The disadvantages of using C++

Overview

• •

Coding for functional verification can be separated into two parts. One is the generic programming part, and the other is the chip testing part. The generic part includes writing structures, functions, and interactions, using techniques such as OOP to manage complexity. The chip testing part includes connecting to the chip, running many threads, and managing random variables.

The generic programming part becomes more and more crucial as the complexity of the hardware to be tested grows. While the problem of connecting to a more complex chip tends to grow only linearly, the overall problem of dealing with this increased complexity grows exponentially.

The authors believe the generic part of programming is served well by C++. That language's features and expressive capabilities far exceed those of any other language currently in widespread use for functional verification.[1] There is already more software written today in C++ than will ever be written for just verification alone. (If one assumes that coders for both domains are equally productive, there are over three million programmers currently, which is and will always will be greater than the number of C++ verification engineers.)

Still, the most important factor is getting our job done. What about the specific issues of connecting to and exercising a chip? Aren't specific languages (such as Vera or "e") better? The answer depends on you, your team, and your project. The verification-specific languages can make it easier to wiggle the chip's "wires," exercise the chip's functionality simultaneously, and vary the configuration of and traffic through the chip. While these tasks are not trivial, the percentage of lines of code required tend to be small.

The tasks that are made simpler by a verification-specific language generally increase linearly with the complexity of the chip. In other words, there are more wires to connect, more independent threads to run, more variables to constrain, and so on.

---

[1] The authors collectively have written in 14 languages and been paid to code in seven of them.

By contrast, it is much more difficult to make the complexity of a chip increase only linearly. So, as a verification system gets bigger, things tend to get out of hand quickly. Our ability to understand a complex verification system is often more important than how we actually connect to the hardware description language (HDL) wires.

Realize that connecting to the chip does not need to be hard in C++. Teal,[2] an open-source C++-to-HDL interface, provides a simple way for C++ code to connect to and interact with a chip. (Teal is provided on the companion CD.)

C++: A Good Verification Language

Several attempts have been made to move verification away from HDLs, such as in the case of Verilog or VHDL.[3] An HDL does a good job of spanning design concepts (called the *register transfer level*, or RTL) down to a few primitives that are used in great numbers to implement a design (called the *gate level*). However, HDLs are not adept at "moving up" in abstraction level to handle modern programming techniques. Specifically, HDLs do not provide for object-oriented concepts. While SystemVerilog makes a step in this direction, it is not clear that such a large span of concepts as SystemVerilog tries to cover can be integrated well into a single language.

The table on the following page lists the pros and cons of various languages suitable for verification.

[2.] Test Environment Abstraction Layer.
[3.] VHSIC (Very High-Speed Integrated Circuit) HDL.

Language	Pros	Cons
Verilog, VHDL	Simple, no extra license required	No class concept, no separation of verification and chip concerns
Cadence Specman "e"	Rich feature set	Proprietary, nonorthogonal language design
OpenVera	C++ "like", better feature set than HDL	Effectively proprietary, interpreted, lacking full OOP support
SystemVerilog	IEEE standard, one tool does all, C interface	Covers all aspects from gates to OOP, implementation compliance is weak, language is large, yet programming feature set is small compared to C++ (no operator overloading, templating, multiple inheritance), weak C interface
SystemC (C++)	Well-known language, open source, does not need a simulator	Big footprint, focus is on modeling, heavy use of templating, coverage and constraint system dominates coding, long compile times, hard to learn
Teal (C++)	Well-known language, good use of C++, open source, few source files	Not a product, no inherent automatic garbage collection
Homegrown PLI/C	Free, well known	Not usually multithreaded, usually called from HDL as a utility function

Many organizations use proprietary C code to implement the verification strategy. These proprietary systems are usually "bolted on" to the HDL code, serving mainly as utility computation functions.

In some sense, open-source Teal and C++ are the natural evolution of this process. Teal allows the verification system to be in control much like the proprietary model that "e," Vera, and SystemVerilog provide. Teal is different in that it is a minimal, yet sufficient, implementation of a modern verification-specific language that uses C++, thus making it easier to learn, use, and maintain.

A Look at Hardware-Verification Languages

Languages dedicated solely to hardware verification did not spring up spontaneously. They were created and are somewhat successful because they have filled a need that languages such as Verilog and VHDL couldn't fill.

As we mentioned earlier, HDLs do a good job of describing hardware concepts, but fall short when it comes to testing hardware—because testing is mainly a software endeavor. Furthermore, HDLs are limited in their ability to manage randomization (which is generally not found in working hardware), and they do not encourage modern programming concepts, such as OOP.

Randomization

It is not simple to manage randomized parameters. It certainly is not clear that we, as an industry, understand enough about managing randomization to have a "best" solution. However, hardware verification languages (HVLs) do address one aspect of randomization: they fill the need for multiple independent streams of random numbers. This feature is necessary for the independent, simultaneous testing of different features of a chip. In addition, the generators are stable across multiple runs, given the same master seed. This is necessary, among other things, to keep the verification system "steady," so that the hardware model can be modified to fix bugs and the exact same test sequence can be run again to confirm that the updated hardware has solved the problem.

The random-number management solution used in HVLs includes a *constraint language*. It is not clear to the authors that this is a benefit. Sure, at some level we have to constrain random numbers to a range (or disjoint ranges), and possibly skew the distribution so that it is nonuniform. However, adding a declarative sublanguage within a procedural verification language is not a clear win. In addition to having to learn an HVL, the application of hierarchical and overlapping constraints is not intuitive.

For example, in one company we used the recommended method of extending a class to add constraints. This is "obvious" in theory, but in a real system one often cannot find, or keep in mind, all the classes and their subclasses. We kept adding constraints that conflicted at run time, and other testers added constraints to a class that many people were already using—even though the added constraints were applicable to only a single test. Finally, we decided that all constraints were to be local to a class, and not in the inherited classes.[4]

Coverage

In addition to using constraints to guide the randomization, HVLs add a *coverage* sublanguage. While coverage is an important concept, the authors are not certain that the industry has a clear solution to apply it. It is a relatively simple matter to collect data, but many questions remain.

- Do you keep the time that the coverage event occurred?

- How do you fold a large coverage range (such as an integer or a real) into coverage bins?

- Most important, what is the relationship between the covered events and the constraints that control the randomization?

This last point is critical. It is the key to the goal of managing randomization in order to meet some coverage goals, so that the system can adjust itself without the need for human intervention. While many papers have been written on automated coverage-driven verification, the authors believe that this is a research question that may not even have a solution for real-world situations.

Object-oriented support

HVLs provide the ability to express verifications systems using OOP techniques. Although each of the HVLs has implemented OOP to some degree or another, this is actually the strong point of C++. While C++ is not the first language to have OOP techniques, it is the most widely used, and has a more extensive and mature OOP implementation than any other current verification language.

[4.] Don't worry if the terms are a bit confusing in this paragraph. They will be explained in the next chapter.

Hardware Verification with C++

Main Benefits of Using C++

· ·

There is a vast difference between the things you think about when coding an RTL design and the things you think about when coding for verification. Using C++ as the verification language helps to clarify this separation of concerns.

C++ is well known, with many books and experts. (A quick web search for "C++" yielded over 77 million references!) The language is well thought out and has been through many revisions, so no sharp edges remain; it has been polished by thousands of engineer-years of practical use. As a result, the features of the language are very orthogonal, meaning that the constructs are consistent regardless of context.

Furthermore, C++ has powerful and proven debugging tools, such as GDB[5] and DDD.[6] Also, because C++ is a superset of C, the existing millions of lines of C code can be easily integrated, including system calls (such as sockets, login, and rsh), shells (such as tcl, Perl, and Python), and databases.

Many companies have C or C++ models of their core algorithms. Common examples include digital signal-processing (DSP) libraries, graphics pipes, or packet-routing algorithms. There are also many public-domain implementations of standard algorithms. Even with specialized algorithms of interfaces, there is often a C-model implementation.

It is odd that hardware verification languages don't work seamlessly with C/C++. These language interfaces support only the integral types (such as `ints` and `chars`), and cannot pass structures or objects to and from a C/C++ model (which is curious, because many HVLs use C++ as the implementation language).

C++ has a wealth of programming features. It supports the concept of interface versus implementation, allowing the user of an interface to concentrate on what the interface can do, not on how it is implemented. C++ also provides operator overloading, templating, namespace, and multiple inheritance, to mention only a few features. Techniques that use

---

[5.] GNU Debugger.
[6.] Data Display Debugger, a popular graphical user interface for command-line debuggers.

such features, to be illustrated in subsequent chapters, allow your verification system to be simple, robust, and adaptable.

Other Benefits of Using C++

In addition to the benefits mentioned above, there are some additional benefits of using C++. These side benefits all revolve around the ability to integrate with the software team—which may or may not be important in your company. With C++ you can enlist members of the software team for your verification effort, either on a short-term or long-term basis. Using C++ for verification also enables the efficient coverification of software and hardware. (The converse is also true.)

As long as a portion of the software team is available as a pool of talent, the flexibility and modularity of C++ makes it possible to use these team members (perhaps with a little help) to aid in writing header files, algorithms, and tests. In addition, the experience of the software team can be used as a "sounding board" for the verification architecture and test plans.

The software team benefits as well. In addition to having a say in the design before it is hardened, they get already debugged header files and key algorithms as well as a better understanding of the chip. There is also the social component of working together, which is beneficial when debugging in the lab.

Coverification is the ability to write and debug some of the system software before the hardware is fabricated. For SoCs, a common technique is to compile a program to binary and load it to the boot RAM in the HDL simulator model. Then during simulation the CPU boots and executes the compiled binary code on the chip.

Instead, by using and executing C++ natively on the simulation host and accessing the chip through bus functional models (BFMs), software algorithms and drivers run orders of magnitude faster. This is because the compiled code runs in the host machine's native language, and drops to the simulator only when a change occurs on a monitored chip signal.

Drawbacks of Using C++

While there are many benefits to using C++, there are, of course, some drawbacks. One drawback is that, by itself, C++ is not a verification solution. Even with Teal or SystemC as the interface between Verilog and C++, additional code must be written.

Another drawback, ironically, is that C++ is a rich language—with the "dangerous" power of expression that this implies. While you will be quite proficient with the language within a few weeks, its expressive capability is immense. Consequently, because it is designed for production software, there are many ways to express the same concept.

Consequently, it can take time to learn how to use C++ effectively, and because there are no marketing teams from large electronic design automation (EDA) companies giving presentations on the design methodology of verification with C++, you have to find your own way. This, by the way, is not necessarily a bad thing.

The purpose of this handbook is to lessen the effects of these drawbacks by providing proven OOP techniques from the software world, and by illustrating, through real examples, how they are applicable to functional verification.

Also, because C++ is an established language, some convenient HDL syntax features cannot be added (for example, "_" or "$" in literals, the "24'hx330z99" notation, or the @ posedge or negedge operators).

In addition, C++ lacks specific verification sublanguage constructs, such as constraint and coverage language declarations. While these can still be added to C++ as libraries, the best implementation is still unclear, because the overall problem is not yet well-understood.

For Further Reading

- *Software Engineering: A Practitioner's Approach*, by Roger S. Pressman, has a great section on the evolution of programming. This handbook also has references to landmark papers and books.

- The SystemC and Testbuilder manuals have discussions on why C++ is good for verification. SystemC information can be found at www.systemc.org, and Testbuilder information can be found at www.testbuilder.net.

- There are several standards for verification and simulation, such as IEEE 1364-1999 (for VHDL), IEEE 1995-2001 (for Verilog), IEEE 1076, P1647 (for the IEEE version of Cadence "e") and P1800 (for SystemVerilog). The web site www.openvera.org provides the OpenVera specification.

OOP, C++, and Verification

C H A P T E R   3

Progress has not followed a straight ascending line, but a spiral with rhythms of progress and retrogression, of evolution and dissolution.

Johann Wolfgang von Goethe

This chapter looks at why and how object-oriented programming was developed, and reflects on why OOP is the right choice for managing the increasing complexity of verification. It then shows how OOP is expressed in C++. The OOP techniques shown in this chapter are used throughout the remainder of this handbook.

Overview

OOP is a programming technique that is often touted as a cure-all for verification. While it is true that OOP is an essential tool in a programmer's toolbox, it is by no means the most important one. One's experience, intelligence, and team environment are far more important to the success of verification than any language feature or technique. That said, OOP is a useful tool for communicating and enforcing design intent for large projects and teams, in addition to being a good way to build adaptable and maintainable code.

This handbook is intended for those with at least some familiarity with OOP. Many verification engineers already have some experience with OOP through languages such as Vera, Specman, SystemVerilog, or SystemC.

The first part of this chapter looks at the history of OOP and why it is well-suited to functional verification. The second part shows how C++ expresses the most common elements of OOP.

For readers with limited experience in OOP, there are several suggestions at the end of this chapter. If you have some, but limited, experience with OOP, or if some time has passed since you used it last, then *don't worry!*

Some of the aspects presented in this and subsequent chapters might seem confusing at first, but the main part of this handbook shows a complete working verification environment. It is the authors' hope and intent that you will "copy and paste" from this environment as well as from the examples provided. This handbook is designed to give you a jump start on using C++ for verification without having to design every class from scratch.

The "basic" OOP techniques expressed in this chapter are the foundation of the techniques used and discussed throughout the remainder of this book.

The Evolution of OOP and C++

OOP techniques, and C++, have been proven to help large programming teams handle code complexity. One key to coping with such complexity is the ability to express the *intent* of the code, thus allowing individual programmers to develop their part of the code more effectively. This understanding of the intent allows programmers to build upon already working code and understand the overall structure into which their code must fit.

Assembly programming: The early days

Programming has changed a lot over the years. It started with the use of assembly language[1] as a way to express a "simple" shorthand notation for the underlying machine language. This simple abstraction allowed programmers to focus on the problem at hand, instead of on the menial and error-prone task of writing each instruction as a hexadecimal or octal integer. Simply put, abstraction allowed an individual programmer to become more productive.

Here is an example of some assembly language:

```
        MOV.W   R3, #100
        MOV.L   R1, #7865DB
loop:   ADDQ.W  R1, #4
        TST.W   R1, R2
        BNZ     loop
```

Procedural languages: The next big step

With the increase in complexity of the problems programmers were asked to handle, procedural languages such as FORTRAN,[2] C, and Pascal were developed. These procedural languages became very popular and allowed individual programmers to become highly productive.

Here is an example of FORTRAN,[3] a common procedural language:

---

[1] The first assembly language was created by Grace Hopper in 1948.
[2] For FORmula TRANslator, created by John W. Backus in 1952.

```
DO 3, LOOP = 1, 10
READ *, MGRADE, AVERAGE
IF (.NOT. (AVERAGE .GT. 6.0 E -1)) THEN
  PRINT *, 'Failing average of ', AVERAGE
  STOP
ELSE
  PRINT *, 'Passing average of', AVERAGE
  AVERAGE = (MGRADE / 1 E 2) + AVERAGE
END IF
3 CONTINUE
```

Interestingly, as the size of the programs grew, the focus of programming switched from the productivity of the individual to the productivity of the larger team. It was found that procedural languages were not well-suited to large programming efforts, because communicating the intent of the code was difficult. OOP, with its ability to build classes upon classes and define interfaces, proved an effective response to this problem.

OOP: Inheritance for functionality

By necessity, OOP developed in stages. The first stage focused on what is often called *data hiding* or *data abstraction*. This is a way to organize large amounts of code into more manageable pieces. With large amounts of procedural code, it became very complicated to keep track of all structures and the procedures that could operate on those structures. It was also hard to expand, in an organized way, upon existing code without directly editing the code—a process that, as we all know, is error prone.

To address these problems, a language called "C with Classes" was developed by Bjarne Stroustrup in 1981. Based on Simula, a simulation language developed in 1967, this new language was recognized as the first language to introduce object-oriented concepts. One of Simula's first uses was as a hardware simulator. "C with Classes" merged these OOP concepts with the C programming language.

"C with Classes" included ways to organize data structures and the functions that operate on those structures, and called this organizational

---

[3.] Okay, you got us—this is actually FORTRAN 77, the "new" FORTRAN (ANSI X3.9, 1978).

concept a *class* (loosely based on Simula's class). The functions, now scoped within a class, are called *methods*. In addition, it included ways for one class to expand upon another through *inheritance* (also from Simula).

Classes allowed for the grouping of code with data, while inheritance allowed a way to express increasingly intricate functionality through the reuse of smaller working modules. This technique is often called *inheritance for functionality*. (Later in this chapter, we'll show how C++ expresses both these features—grouping into classes and reuse through inheritance—in more detail.) This new approach was sort of like the Industrial Revolution of the programming world, increasing team productivity by an order of magnitude.

So "C with Classes" helped improve the productivity of programming teams, by helping to organize the code in layers—with one layer inheriting from, and enhancing upon, a lower layer. This meant that the code could now be "reasoned about." With this technique, changes and bug fixes are made only to the appropriate module, without the changes echoing, or propagating undesirably, throughout all of the code.

Furthermore, as code was structured into layers through hierarchy trees, several patterns became visible. For example, it became clear that certain layers were not involved with manipulating the data (in the classes) directly, but rather with ordering, structuring, and tracking events.

These framework layers became more and more important to understanding the system. To get a large program to be "reasonable," more and more standard infrastructure was needed. These framework layers had no "interest" in how the actual data were manipulated; rather, the important feature was that now the data could be assumed to be manipulated in predefined ways.

As an example, as long as each class in a particular framework layer had a `start()` or a `randomize()` function, working with classes of that type was reasonable. As these framework layers were written, it became clear that they could be generalized as long as each class followed the rules for that type of "component." The problem was that "C with Classes" provided no way of enforcing these rules.

OOP: Inheritance for interface

What was needed was a way to express the rules that a class had to follow in order to "fit in." The solution, known as *virtualization*, was included from day one in the language that would become known as C++. With virtualization one could define classes called *abstract base classes*, which simply expressed the *interface* a component must conform to in order to fit into the larger system.

Each developer of the actual classes that fit in a particular structure would then inherit from this abstract base class, and implement the details for how a particular function should be implemented for the problem at hand. This technique of defining the interface through virtualization, often called *inheritance for interface*, is frequently used in OOP-based projects.

The clever thing is that now one could write the code for the framework layer using abstract base classes. This not only allowed the framework to be implemented concurrently with the data-based classes, but it also allowed the framework layer to be developed in a much more generic way. This virtualization of base classes has proven to be a powerful technique for creating and maintaining large and complex systems.

OOP: Templates

Virtual base classes are a great development for generic programming, and are a very powerful tool for expressing how an environment or program should be structured. However, there are certain problems that are even more generic in nature, such as linked lists or dynamic arrays. For example, verification code often maintains arrays of generated data and arrays of the data that have been received.

For these container-based problems, the point of the base class is algorithmic and independent of the data. Consequently, it would be clumsy to have to inherit from a base class just to use the algorithm. Instead, a technique called *templating* was developed. A template is, not surprisingly, a reusable pattern for code, rather than an implementation. (Templates are discussed later in this chapter.) The idea was to abstract the algorithm away from the data it operated on. In effect, this is also a framework of sorts.

C++: The standard template library

The most common grouping of templates is the Standard Template Library for C++ (commonly known as "the STL").[4] The STL is a large selection of container and algorithmic templates that solve a lot of common programming problems, such as the storing, ordering, transforming, and managing of groups of data. The STL is an indispensable tool for day-to-day work, and has proven to be more efficient (in time and space) than what most programmers can write.

C++ contains many more features, but its key strength is that it allows for a programmer's intent to be expressed through the ability to build upon known working classes. C++ is built upon layers of work in earlier languages.

The Evolution of Functional Verification

Verification through inspection

There are similarities with the development of OOP and that of functional verification, and while hardware verification is a younger field than software programming, it has (not surprisingly) followed a similar path.

As readers of this handbook surely know, functional verification has come a long way from its recent humble beginning as a (mostly manual) process of verifying simulation waveforms. From there, it evolved into "golden" files; a current simulation run was compared to a known-to-be-good result file—the golden file. For this technique to work it required fixed stimuli, often provided in simple text format. Golden files were an acceptable technique for small designs, where the complete design could be tested exhaustively through a few simulation runs.

[4.] Invented by Alex Stepanov in 1992.

Verification through randomness

The simple technique of using golden files became impossible to use as the size of the hardware being tested grew both in size and complexity, so other techniques were needed. For larger projects it was no longer possible to test the "state space" of a chip completely. To do so would require an unobtainable amount of computer time, even on the fastest machines. To address the reality that the chips being developed could no longer be tested exhaustively, random testing was introduced. Using randomness in tests changes the input stimuli every time a test is run. The goal is to cover as much of the state space as possible through ongoing regression runs.

Unfortunately, several problems were found in using randomness with current hardware description languages (such as Verilog or VHDL). To begin with, the result checking became more complex as golden files could no longer be used (because the input stimuli changed for each run). This meant that verification models that could predict the outcome from any set of input stimuli were needed. Writing these models has become a major task for the verification projects of today.

However, this technique also posed other problems. It was discovered that using randomness was a tricky thing. If you use random stimuli and your testing fails because of a hardware bug, then you later would want to rerun the exact same sequence of stimuli to verify that the bug has been solved. This is easier said than done.

You can either record all the stimuli that generated the test run, then use some mechanism to replay the stimuli later; alternatively, you can track the "seed" from which the (pseudo) random generator starts and then pass that number into your next simulation run.

Both techniques can be problematic, because storing all the generated stimuli requires a lot of disk space and directory infrastructure, and because controlling randomness through a seed requires good control over your "random" generator.

The current most common solution to this problem is to control and store the "random" seed, then use it to replay a given stimuli sequence over and over.

The emergence of
hardware verification languages

We can see that controlling the generation of random stimuli requires many things. We need verification models that can predict results from any given set of stimuli. We also need control over how the random generator works, to be able to replay a given stimuli sequence. It was found that using HDL languages, such as Verilog and VHDL, was difficult with respect both to writing high-level models quickly and controlling randomness. In Verilog, for example, you couldn't control the random seed back in 1987.

As a result, people started looking at other languages for verification. The natural first step was connecting C to Verilog, but soon languages such as "e" and Vera were introduced. These languages made it easier to do random testing, in turn making it possible to test much larger chips.

OOP: A current trend in
hardware verification languages

The problems we are facing today in verification are similar to the problems software faced when OOP was adopted. We now have to deal with very large amounts of code and multitudes of modules, all of which must be compiled, instantiated, controlled, randomized, and run. This is not an easy task, and we spend more and more time solving these basic framework problems.

It seems clear that adopting OOP techniques should help make these problems more manageable. Unfortunately, there are still not enough people in the field of verification who have sufficient experience and understanding of how to develop an appropriate OOP infrastructure.

Engineers in our field are just starting to adopt OOP techniques, and as we are early in this process, they are taking many different approaches. Vera and SystemVerilog offer one approach, by trying to marry many different concepts into one complete verification language.

SystemVerilog, for example, covers many more aspects beyond just OOP, including an RTL language, a temporal assertion language, a constraint language, and even a code coverage language.

Contrast this with the development of C++, where there was always an effort to keep the language small. Most of the syntax of C++ is still compatible with C. New techniques were carefully explored before they were introduced, as can be seen, for example, in virtualization. The concept of virtualization was known before "C with Classes," but it was left out of the language until it could be implemented well.

OOP: Problems with the current approach

OOP is a multifaceted concept, and as such requires flexibility of the programming language to work well. Therefore, C++, to encourage appropriate OOP techniques, implements many powerful features (such as templates or multiple inheritance). These features were added not frivolously, but instead to support flexibility. Removing too many of these features from a language would make using OOP clumsy or impossible.

A great feature of C++ is that the powerful STL library is declared directly in C++ (through templates). In addition, the popularity of C++ has led to implementations available as open-source code. This means that new templates can be added or current ones enhanced, as needed for a particular problem. This has created a large amount of code available as open source.[5]

OOP: A possible next step

The field of verification is young; not long ago we were staring at waveforms on a screen. By using modern verification languages we have developed the field into something better. However, today we are facing even harder problems, one of which is the issue of the *framework*. To do a job that is increasingly complex, we need a framework for how our verification environment is interconnected. This is no longer an easy thing to achieve. In this handbook we show many techniques for how to manage this and other problems. We also introduce an open-source verification framework, called *Truss,* that collects our best experience in OOP into a working environment.

[5.] The site www.sourceforge.net has over one-hundred thousand open-source projects.

Hardware Verification with C++

It is our belief that if enough people adopt a powerful open-source infrastructure, many great innovations will result. The problem we face today cannot be solved by the features of individual languages alone; rather, we need an agreed-upon framework. Even if this framework were modified by each team, it still provides the opportunity for best practices to evolve. This handbook, and the accompanying CD, is our attempt to start the discussion.

However, we are getting ahead of ourselves, so before we dive into the practical problem of verification, let's look at how C++ expresses OOP techniques.

OOP Using C++

This section shows how C++ expresses the concepts of OOP described above. It describes some of the techniques we use to build a successful verification environment in later chapters. For engineers experienced with other OOP languages such as Vera or e, this chapter can serve as a way to map concepts from one language to another.

Data abstraction through classes

Using classes to express data abstraction is an important technique in building large verification systems. Data abstraction, by grouping the data and the operations together, allows engineers to reason about the code.

We will look at a Direct Memory Access (DMA) descriptor class to show how a class can be constructed, then improve upon it by using inheritance.

A DMA descriptor example

DMA is a common hardware feature for transferring data from one memory location to another without putting a load on the CPU. In this example, we verify a DMA chip that accepts DMA descriptors, puts them into an on-chip memory array, and then executes them. Each descriptor has a source and destination memory address, as well as the number of bytes (called "length") to transfer.

In the verification environment, a DMA descriptor is represented by a small class. The DMA generator is then responsible for building, or *instantiating*, DMA descriptors and "pushing" them to the chip and to the checker.

The following code describes the DMA descriptor class:

```
class descriptor{
  public:
    //Constructor and Destructor
    descriptor(uint32 src, uint32 dest,
            uint32 length, uint32 status);
    virtual ~descriptor();

    //Methods and Variables
    uint32 source_address;
    uint32 destination_address;
    uint32 length;
    uint32 status;
    const uint32 verif_id;
    virtual void print();

    //Operator Overload
    virtual descriptor&  operator== (const descriptor&);

  private:
    //Copy Constructor
    descriptor(const descriptor &);
    descriptor(); //No implementation; don't call method
    descriptor& operator= (const descriptor&);
};
```

The descriptor class is divided into the `public:` and the `private:` sections, which declare the access control level for the main program. The `public:` section is the "user interface," which in turn consists of three parts: the Constructor and Destructor, Method and Variables, and Operator Overload sections. Let's look at what these sections mean.

Access control

The keywords `public:` and `private:`, as used in the `descriptor` class, are C++ *access control* labels. They indicate how methods and variables following the statements can be used. A *public* label indicates that the methods and variables following are accessible by any code that has access to an instance of the class. A *private* label indicates that only the code inside the class itself can access the variable or method.

Access control is needed to help separate the user interface from both the internal methods and the data needed to implement the class. Consequently, the *public* section of a class declaration is the "user interface." These are the interesting methods and variables to look at when you want to *use* a new class. When you *implement* a class, on the other hand, you also need a space to store the "state" of your class between method calls.

Implementing a class is similar to implementing a state machine, where each method call changes the state of the state machine (that is, modifies the data members of the class). This "change of state" must be recorded somehow. Variables for tracking the state as well as intermediate methods should be put in the *private* scope, not only to indicate to users that they shouldn't focus on these methods and variables, but also to protect these variables from accidentally being modified. When a class is instantiated, only the public methods and calls can be accessed. Trying to access private scope results in an error during compilation. This is an example of how language enforces the "intent" of the class.

Enforcing intent can (and should) go beyond protecting state variables. Many times, when one creates a class certain ways of instantiating or copying the class are not possible to implement. For example, instead of printing an error message during run-time, when the code calls unsupported methods, one should declare those methods to be in private scope, so that a compile error occurs instead. In the *descriptor* class, access control labels are used to communicate these intentions by keeping the copy constructor in private scope.[6]

---

6. Note that C++ does resolution first, then access checking. This is extremely important, as it ensures the code behaves well when the access control is changed. Note also that the current HVLs do not follow the rule of applying resolution first, then scope. This problem is discussed in detail in the book *The Design and Evolution of C++*.

C++ actually provides three levels of access control. In addition to *public* and *private,* there is also *protected.* Protected indicates a private variable or method that can be modified through inheritance. Public, private, and protected can be used to express and enforce the intent of the class clearly.

Constructors and destructors

Constructors and *destructors* are the methods called when a class is instantiated and goes out of scope, respectively. Constructors are used to initialize member variables, reserve memory, and initialize the class. Destructors, conversely, serve to free allocated memory and perform any cleanup work. A constructor is defined as a method with the same name as the class it defines—in our example it's called `descriptor(...)`— while destructors are named the same but with a preceding tilde (~), as in `~descriptor()`.

One thing to remember is that C++ has several types of constructors, depending upon how a class is instantiated. These can be described as the *constructor*, the *copy constructor*, and the *default method.* C++ even goes so far as to create the copy and default constructor automatically for any class, as long as it's possible to do so.

Consider our descriptor class for a second. The class could be instantiated as follows:

```
descriptor descriptor1(source_addr, destination_addr,
                       source, length);
descriptor descriptor2(descriptor1);
//line will fail to compile - method in private scope!
descriptor descriptor3; //fails!
```

The first line calls the constructor method as declared in the code above, passing in variables as necessary. The second line calls the copy constructor, meaning that `descriptor2` will be created as an exact copy of `descriptor1`. The third line calls the default constructor, so that a descriptor is constructed with "default" values, if possible. Each line is valid C++, but this might not be what the class writer intended.

In our descriptor class example, and frequently in verification, it might not be a good idea to allow for the default or copy constructor. The default constructor in the descriptor example doesn't make sense, because the user needs to provide guidance regarding the source, destination, and number of bytes to copy. The copy constructor doesn't make sense either, because of a verification trick we have decided to use in this example.

In the descriptor class there is field called `verif_id`. Our trick is that a unique number is passed for each DMA transaction generated. This number is passed through the chip to be verified and enables the checking code to "understand" which DMA transaction completed, so that the result in hardware can be compared to the stimuli. However, if objects are copied, then descriptors with the same ID are passed through, making verification and debugging hard.

To stop this unwanted behavior and enforce intent, the copy and default constructors in our example have been moved to private space. This means that if the copy constructor is called by mistake, a compile error will occur.

When writing a class, try to express the intent of the class so that an unintended use of your class generates a compile error. Though annoying, compile errors are much easier to understand than run-time errors. Similarly, when you get an "unexpected" compiler error from a new class, don't see it as an annoyance, but rather realize that it might be that the person who wrote the class is trying to tell you something.

Member methods and variables

In the descriptor class example, a few member variables and member methods are declared. The member variables are simply integers for the fields of the DMA transaction. The `verif_id` variable is declared as a `const`. This means that the variable, once initialized (by the constructor), is constant and can't be changed again. This is a valuable technique for variables that you don't want changed. It is important that `verif_id` remain unchanged throughout the simulation if the predictor is to work. The `print()` method will simply print all the current fields of the projects. Member variables and methods can be accessed like this:

```
descriptor my_descriptor(0x68000, 0x20500, 39);

my_descriptor.source_address = 0x586;
my_descriptor.print();
```

Operator overloading

In the descriptor class example, there is also one operator method. This is called *operator overloading*. Operator overloading is a powerful concept in C++. It allows you to define how any arithmetic operation (such as =, ++) acts upon a class. With C++, classes and enumerations can be indistinguishable from built-in types. This is essential when you want to define a class or `enum` that a user can intuitively apply in binary operations (such as <, ==, !=, +, and *), as well as in unary operations (such as ++, --, and !). You can also provide specific behavior for the operators `new()`, `delete()`, `->()`, and `operator()`, which provide a wide range of powerful capabilities. (The STL has some good examples of this.)

In the DMA example the equality operator is overloaded because when two DMA descriptors are compared by the checker, we don't want to compare the `verif_id` field—we already know that it is unique for each DMA descriptor. Consequently, in the implementation of the `operator==()`, we can simply compare all other fields.

Operator overloading is a powerful technique to make classes behave like built-in types—but beware operator overloading for classes that don't logically have operator functionality. It is not clear, for example, what `operator--()` would do for the DMA class above.

Inheritance for functionality

By using class inheritance, you can create larger and larger functional blocks, building upon existing functionality. By inheriting from another class, you are saying, "I want to start from the functionality of an existing class and expand upon or change it with the features I define in my new class."

Consider our DMA project again. In the first generation of the product (as described above), the chip would simply store DMA descriptors in an array and signal when the array was full. For the second generation, this has been improved and the chip now implements a linked list, storing each descriptor in off-chip memory.

To enable this functionality, a pointer field must be added for each descriptor. As a technique, the pointer can be set to null to stop the chip from processing, or it can be set to the first descriptor to implement a ring.

Instead of copying and editing the descriptor class, we can simply inherit from and expand upon the base descriptor class. This is called *inheritance for functionality*.

To create our new, fancy `linked_list_descriptor` class, we would declare it like this:

```
class linked_list_descriptor : public descriptor {
  public:
    linked_list_descriptor(uint32 src, uint32 dest,
                           uint32 length, uint32 next);
    virtual ~linked_list_descriptor();
    virtual void print();
    linked_list_descriptor& operator=
                    (const linked_list_descriptor&);
  private:
    uint32 next_descriptor; //Pointer to next DMA memory
    linked_list_descriptor
                    (const linked_list_descriptor&);
};
```

The `:public descriptor` from the first line of the class states that the `linked_list_descriptor` class inherits from the descriptor class. Now the `linked_list_descriptor` class has all the functionality of the original descriptor class, and adds in (among other things) the `next_descriptor` variable.

There is another thing to note about the `:public descriptor` declaration in the class declaration. This provides access control, in a way similar to that described for member variables above, and indicates that the DMA class is directly available to users. Private and protected inheritance is also allowed. (This is beyond the scope of this chapter, but the handbook provides examples of private and protected inheritance in later chapters.)

Inheritance for interface

As described previously in this chapter, in the section "The Evolution of OOP and C++," *inheritance for interface* means using a base class (with virtual methods) to describe the framework of a system. With this technique an *abstract base class* is used to define an interface that classes inherited from it must implement. This base class defines the methods as *virtual*, which indicates they can (and sometime must) be defined by the class inheriting from the base class.

With this technique, standard interfaces for similar, but different, components can be used. This is very useful in creating a verification framework, because in a large verification environment you must keep track of a large number of components—for example, verification components (such as bus functional models, generators, checkers, and monitors), and test components. By defining an interface to which each type of component must conform, the ability to reason about the environment increases.

(It should be noted that defining appropriate abstract base classes is not easy. Overly complicated or overly generic base classes tend to make the problem of verification harder. In later chapters we'll talk about the trade-offs.)

For an example, let's consider building an abstract base class for a bus functional model (BFM) for a verification project. It has been decided that all verification components need certain *phases* (expressed as method calls), including `randomize()`, `out_of_reset()`, `start()`, and `final_report()`.

These methods ensure that a verification component is randomized, has time to program its part of the chip [in `out_of_reset()`], starts up any threads needed to run, and has a way to print its status once the simulation is done. This can be done by creating an abstract base class from which all actual drivers inherit. An abstract base class in our example could look something like the following:

```
class verification_base {
public:
  virtual void out_of_reset(){/*do nothing */};
  virtual void randomize(){/*do nothing */};
  virtual void start()=0;
  virtual void final_report()=0;
}
```

What makes the class abstract is the fact that its member methods are declared by means of the keyword `virtual` (and at least one virtual method is pure, as described below). This class declares how the verification system framework expects any verification components to behave, by enforcing that all verification component have at least these methods.

In C++ there are two types of virtual function: *virtual* and *pure virtual*. In our example above, the first two methods are virtual, the last two are *pure* virtual. Virtual functions have a "default" implementation (in our case they do nothing), while pure virtual functions have no implementation and are indicated by "=0". A pure virtual function is one that a derived class is obliged to implement. For virtual methods, the original method is used if no same-named method is declared.

Consider an Ethernet driver, which is inherited from the class `verification_base`.

```
class ethernet_driver : public verification base {
  public:
    void out_of_reset() {set_up_dut();}
    void start();
    void final_report();
  }
```

In our Ethernet driver, `randomize()` is not declared, so the default method specified in `verification_base` is used.

This technique of using inheritance for interface is very important for creating a flexible, yet reasonable, verification structure. In our verification domain, many objects must be initialized through many phases, synchronized, and run. This is not an easy task for anything but the smallest projects. However, by using inheritance for interface and virtual methods, once can create a powerful and flexible verification environment.

Templating

One of the features that make C++ such a rich language is its template library, the STL. The STL is part of the ISO C++ standard and is shipped with every implementation. Many useful day-to-day features, such as handling "dynamic arrays" (considered "new" in dedicated verification languages such as SystemVerilog), were long ago introduced as templates in C++. In addition, C++ implements these features as normal code, not as a language "feature," so that new templates can be developed by the community.

There are many different types of dynamic arrays, often called *container classes*, with the most common ones being *vector*, *dequeue*, and *list*. Dynamic arrays extend the C pointer concept and package the concepts in well-known abstractions, such as queues, lists, and so on. These containers can hold a variable number of elements, and the user can add, subtract, reorder, and modify the elements. The `std::vector` class, for example, can be seen as a dynamic array, where any element can be addressed directly like an array (for example, `myVector[7] = 0x39A`).

Using these classes greatly increases the productivity of the programmer, while still maintaining speed of execution (STL containers are very efficient in time and space). The beautiful fact is that an element in a container can be almost anything—from built-in types (such as `int` or `char`) to user-defined classes or pointers to them.

A goal of the STL was to provide generic container classes to help broaden the scope and appeal of C++. The STL has been very successful in providing a large number of well-defined, simple-to-use, container classes and functions that operate on containers.

If you are not already familiar with the STL, the authors strongly recommend that you take a closer look at it. The STL is very useful for storing and sorting data, something rather common in our world. (See the end of this chapter for a few books on STL.)

The C++ template library is vast, debugged, efficient, powerful, and useful.

A word about user-defined templates

In C++ you can write your own templates, which can be useful for expressing certain type of intent. However, the authors recommend that you not write your own templates until you are very comfortable with C++. Instead, rely on the STL and other proven templates. For verification, there are few problems that are better solved with user-defined templates.

There are two main parts to templating: defining the template, and using the template. The difficulty is not symmetric, and it is far easier to use a well-defined template (such as the ones found in the STL) than to build one. Unfortunately, it can also range from hard to nearly impossible to use a vague or badly defined template. For this reason, the authors recommend against writing many templates for verification. That said, there are times when creating a template is appropriate.

Namespaces

It happens in every large verification project. You try to link all compiled files together and run into conflicting variable names; it seems that there is always more then one module called "generator" or "driver." It's frustrating, because now you have to go back and rename the conflicting classes and files. Furthermore, the "new rules to follow" probably becomes "you must insert interface name before variable name," so you end up with `uart_generator` and `ethernet_driver`.

But how do you deal with code from Intellectual Property (IP) vendors? How do you know what names *they* use?

C++ has a solution to this common problem: the use of *namespaces*. A namespace is the placement of related classes and global functions in a logical group. For example, if you are testing an Ethernet interface, all your classes and components might go into the `ethernet` namespace. If you testing a UART interface, consider using the `uart` namespace, and so on.

A namespace is simply declared as follows:

```
namespace pci_x {
  class master {
    ...
  }
}
```

Any class or variable wrapped inside the brackets is now in the `pci_x` namespace. When you later want to instantiate a `pci_x` master, you simply declare what namespace you are using and what module you want.

Note, for example, the following:

```
pci_x::master my_master(default_master);
```

The `pci_x::` indicates that you want to instantiate the master class from the `pci_x` namespace. If you are using a lot of components from a certain namespace, you can declare that you want to have access to that namespace throughout your file by means of the keyword `using`, as follows:

```
using namespace pci_x;
```

From that point on, you have access to all components in the `pci_x` namespace.

However, be careful about putting a `using` clause in a header file. *This is almost always a mistake.* The reason is that the `using` clause has now been added to every file that directly or indirectly includes this header file. So, the fact that some code was in a namespace is now lost to code that includes this header file. The authors have experienced where this caused the very name collisions that namespaces were designed to avoid.

There is another nice feature about namespaces that is worth mentioning. You can continue adding to a namespace by simply "redeclaring" it. At a later point, when you are ready to add some new code to an existing namespace, you can simply put the same namespace tag on the new code.

For example, adding to the `pci_x` namespace we talked about above, you would say the following in any header file (a header file is, by convention, a file that just provides the interface of a class):

```
namespace pci_x {
  class slave {
    ...
  }
}
```

Now, by including both header files, the `pci_x` namespace contains both a master and a slave.

Summary

This chapter wove together three themes: the evolution of C++, the evolution of verification, and the way C++ expresses OOP features.

We spent some time looking at the class declaration, with its accessor labels, constructors and destructors, data members, and operator overloading.

We then took a look at inheritance and its two main techniques: implementation reuse and interface specification.

We covered templating next, as it is a valuable addition to the techniques we can use. We also discussed the C++ template library STL, and how it is easier to use templates than create them.

Finally, we discussed namespaces as a way to avoid name collisions.

For Further Reading

- *The Design and Evolution of C++,* by Bjarne Stroustrup (the father of C++), is a great read for those interested in how the features of C++ came into being.

- *The C++ Programming Language*, also by Bjarne Stroustrup, is the best reference book on C++, including the STL.

- *Accelerated C++: Practical Programming by Example,* by Andrew Koenig and Barbara E. Moo, is a great book by the experts for the rest of us.

- *C++ Primer (4th Edition),* by Stanley B. Lippman, is another classic C++ book.

- A good book for getting better at C++ once you know the basics is *Effective C++: 50 Specific Ways to Improve Your Programs and Designs*, by Scott Meyers.

- Also by Scott Meyers, *Effective STL: 50 Specific Ways to Improve Your Use of the Standard Template Library,* is similarly good for learning more about STL.

- The web has a wealth of information about C++. The authors feel the best place to start is C++ creator's home page, at http://public.research.att.com/~bs.

A Layered Approach

It is tempting, if the only tool you have is a hammer, to treat everything as if it were a nail.

Abraham Maslow

For longer than we know, humans have organized themselves into layers. From the family and tribe all the way up to national governments, we have created roles and responsibilities. Closer to the hardware domain, both VHDL and Verilog also use a layering concept, employing entities or modules to break up a task. The software domain uses the related concepts of procedures (*methods*) and data structures (*classes*). A reason we humans make layers, with associated roles and responsibilities, is to simplify our lives.

This chapter looks at how using layers can organize the task of verifying a chip. We look at a generic chip, albeit one with a "System-on-a-Chip" bias, and come up with a set of standard, well-defined layers, roles, and responsibilities. We leave this chapter with definitions of standard verification layers and detailed diagrams of functional "boxes" and how they are interconnected. Part II of this handbook will show a fully implemented C++ environment that uses this approach. Part III will talk about general C++ techniques for implementing these classes and connections. These techniques, applicable to most of the languages used for verification, express the reasoning behind the layered approach discussed below.

Overview

Throughout this chapter, little distinction is made among architecture, design, and coding. This is because these activities are interrelated, and occur at most stages in a project. Also, even with the initial architectural efforts, you should have a plausible implementation in mind; otherwise, the architecture may create problems when you are coding.

At many layers, verification environments tend to have the same set of problems. The essence of this chapter is to show how these common problems can lead to common solutions. By reusing solutions, the team can be more productive.

Specifically, this chapter covers the following topics:

- The importance of code layers, roles, and responsibilities
- How to go from a whiteboard verification system to classes and interconnects, using a standard framework
- Some common components, roles, and responsibilities of a verification system

There are many successful hardware products. Because success demands more success, the hardware produced in the next revision of a product will be more complex than the current version. In addition, the sales staff wants the product in the shortest possible time. The three competing factors of quality, functionality, and time to market create stress on the verification team. You are expected to produce more in less time—*and* with increased quality.

So how do you do that? You could add members to your team. While it is certainly true that there is an appropriate number of people for every task, adding people creates several issues. One is the need for increased communication; adding a team member increases the need for each team member to interconnect with the rest of the team, decreasing productivity.

Another issue with adding members is team dynamics; each time new people are added to a team, it takes time for the team to become fully productive again. Finally, there is the fact that a well-integrated team can outperform an average team by a huge margin.

Okay, so adding people is a difficult way to build a quality verification system faster. The authors believe that a good way to do more quality work in less time is to increase productivity. As humans have done in the past, productivity can be increased by using layers. Now, we are not saying a government is a superefficient operation, but rather that a small team can be more efficient if the verification system is divided into layers. In addition, the resulting system is more likely to be simpler and able to be "warmed-over" for the next project.

> *A major tenet of this handbook is that the most productive individuals and teams use a layered approach.*

By using layers to separate the tasks of verification, common techniques and solutions can be seen. This allows the team to build up a library of standard solutions to common problems. Each of these solutions can be given a name (sometimes as a *base class*), along with a defined role and responsibility.

In this chapter we use layers to create a verification system. Starting with a whiteboard block diagram, we define layers, roles, and responsibilities and (in theory) arrive at a well-designed system. This technique is used to show how to move a verification system quickly from a whiteboard block diagram to classes and functional code.

We do not talk much about C++ in this chapter, because the technique of using layers is applicable to almost any language. Part II of this handbook shows specific implementations in C++.

A Whiteboard Drawing

Most verification systems start on a whiteboard or something similar. Some engineers get together and discuss how they are going to test some part of the chip or maybe the entire new product. This initial effort results in an understandable and "clean" block diagram. However, transforming this whiteboard sketch to a similarly clear code architecture and implementation is difficult. This chapter outlines a layered approach to this transformation.

Note that the layering process occurs in one form or another at many levels of a verification system, from the full system level down to individual functional blocks. In addition, the classes and code are constantly refined and modified as the project progresses, so the use of these techniques is both fractal and recursive. This section focuses on this OOP process at the outermost level—in other words, from a system perspective.

A top-level whiteboard drawing might look something like this:

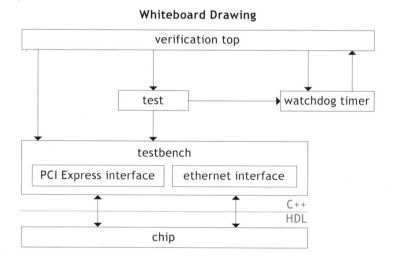

Whiteboard Drawing

- The *verification top* block is responsible for instantiating and ordering the events of other components.

- The *test* block is responsible for setting up, controlling, and synchronizing each interface component.

- The *watchdog timer* is a time-out module that ends the simulation should something unexpected occur that would make a test run forever.

- The *testbench* block is responsible for instantiating each *interface* block. The PCI Express and Ethernet interface components here are important, because they transform test commands into bus transactions on the chip. They include methods for data generation and randomization, as well as drivers and monitors, as will be discussed in greater detail below.

An "ends-in" approach

So where should you start after the first whiteboard drawing is done? A common approach involves starting at the lowest level of abstraction and coding the next layer up, continuing upward until every layer—including the test layer—has been designed. This is called the "bottom-up" approach.

However, there is another approach. This alternate approach still starts with the lowest layer of abstraction (because this is the best-defined layer), but then builds the top layer next and saves the middle layers for last. This approach is called an "ends-in" design, because you start by working on the interface layer and the top-most layer, and then build your way inward. This is the authors' preferred approach, because it maximizes what you already know. You know what the interfaces of the chip are (or at least most of them). You have an idea what a standard test looks like. You can therefore build these layers so that they are "reasonable" (that is, others—on the team or off—can reason intelligently about them). It's then an engineering effort to make trade-offs between complexity and adaptability, in order to connect the ends together into a system that is reasonable at all layers. This is not simple, and it requires a lot of experience, but the next part of this handbook contains a working example of how to do it. Part III of the handbook discusses techniques to evaluate the trade-offs.

Refining the whiteboard blocks

It would be tempting to define a class for each block in the whiteboard drawing shown in the preceding section. While it is possible to do so, this is not a good solution, because each block, especially an interface block, contains too much functionality to fit well into a single class. Having classes that are too large leads to a brittle and often complex design that is not adaptable, or even maintainable.

Instead, it is a good idea to look closer at each block and define another set of layers. This makes sense, because most blocks can have several well-defined abstraction layers. This is what the rest of the chapter will address. Each major section below introduces the general roles and responsibilities of a block or abstraction level. The sections even get a bit more specific, suggesting common names for classes at each level. Some of these names are already common in our industry.

The "Common-Currency" Components

The first step in the transformation from a whiteboard diagram to code is to focus on the chip interfaces. The authors call the set of resulting classes the *interface* layer. In the whiteboard drawing above, there was a PCI Express interface and an Ethernet interface. Because these interface blocks cover a lot of functionality, there needs to be a set of classes for each block, as discussed below.

The resulting interface classes are an example of a design pattern that the authors call the *common currency* of a verification system, because they are used so frequently. In fact, the interface classes are an implementation of the common-currency pattern, because the chip can have most of the interfaces running for every test. Put simply, common currency can be considered a concept or pattern, of which interface classes (such as PCI Express and Ethernet) are specific instances. These classes are the "money" of the verification system's "economy." Every team member should be able to identify the currency—in other words, the roles and responsibilities of the various interfaces of your chip.

There are many ways to identify a common-currency class in an OOP language. One way is to have the class inherit from a common-currency base, such as `class pci_express_monitor : public monitor`. In this case the `monitor` base class has a set of methods that `pci_express`[1] is expected to implement. Another way is to use a naming convention, such as `class ethernet_monitor`. Note the absence of the base class. While the "monitor-ness" of the class is not enforced by the compiler, you can bet the team will have expectations about what this class does.

Sometimes this naming-convention approach is best if the base class has no methods, or has just light-weight ones such as `start()`, `stop()`, and `report()`. The art of deciding what is a class, a convention, or a base class is up to you. The OOP part of the handbook discusses the various options and trade-offs.

The Interface Layer in Detail

As mentioned above, each interface component can be divided into more layers and classes. This promotes adaptability and makes sense, because an interface component straddles abstraction layers; at the highest abstraction layer it consumes transactions, and at the lowest it wiggles wires.

The approach used by the authors is to break each interface into three sublayers. The lower a class is in the interface layer, the more the chip details that are handled. This layering process is a technique to manage complexity by allowing higher-level code (such as generators or monitors) to describe the problem in a more abstract way, thus providing a simpler interface to the tests and making them both clearer and more portable.

The following figure shows how the interface layer is in turn broken down.

[1.] Or any other class that extends the monitor base class.

Interface Classes

There are three abstraction layers. The *transaction* layer consists of fairly high-level classes, such as generators and checkers. The next layer down in detail is the *agent* layer. This is the layer that implements the connection to the low-level "wire wigglers" and is responsible for keeping track of data transactions. The lowest layer is the *wire* layer, in that the objects in this layer drive and sense the chip wires. Let's look at the wire layer first.

The wire layer

The most detailed layer of the common-currency classes is where the monitor, drivers, and bus functional models (BFMs) exist.[2] This is shown in the highlighted section of the following figure.

[2.] With multilayered protocols, the fractal nature of a layer must be considered. Depending on the test to be run, there will be monitors, drivers, and so on at each level of the protocol.

Interface Classes - Wire Layer

In this handbook, monitors and drivers are considered one-way connections,[3] while BFMs are considered two-way connections. These classes are generally the only ones that drive or sense the wires.

Note that because we are using C++, the monitors, drivers, and BFMs cannot directly modify the HDL values. Instead, a shim layer—a *C++-to-HDL gasket*—is used.

The wire-layer classes are complex and have a broad interface. In other words, they have lots of methods, encompassing everything you want to exercise. The classes are extremely portable, because, by definition, the protocol is between chips. If, on another project, you have that same interface, the monitor/driver/BFM should be easily adaptable.

The wire-layer classes have only a simple procedural interface. Their role is to take method and procedure calls and execute the wire dance that is specified by protocol. These classes are responsible for the mapping between a method call and wire-change sequencing. Whether, and in what order, these methods are called is the concern of the next layer.

[3.] A driver sends data and a monitor receives data. Note that the driver or monitor may both drive and sense wires to do this function.

The agent layer

The next layer up is called the *agent* layer, as shown here:

Interface Classes – Agent Layer

The agent layer is responsible for using various wire-layer classes to implement the upper-layer requests. It is called the agent layer because it acts as a go-between for two relatively well-defined components. Commonly, the classes in this layer add some sort of queue, for data or control actions, depending on what the upper layers generate or check.

This layer may also have several implementations. For example, many chips have multiple ways to send the same data. There could be register, FIFO, and DMA ways to interact with a chip. You could have three different connection classes for each of these methods. The test could randomly pick which method to use, and would still look the same.

Because the agent layer is the action layer's view of the chip, it also implements the connection policy. For example, you could use a simple direct connection, thus forcing the generator and driver to act in tandem. Alternatively, you could implement a multipoint connection, using events or other broadcast mechanisms, to connect several drivers to a single generator. The same concepts can be used for the monitor-to-checker connections.

• • • • • • •

The transaction layer

The uppermost layer is called the *transaction* layer, as shown in the following figure.

Interface Classes – Transaction Layer

The transaction layer uses the previously discussed layers to exercise some interface or feature of the chip and validate the response. The exercising (or driving) part of this layer is called a *generator*. The response validating (or receiving) part is called a *checker*. Note that there may be more than one generator or checker if different types of traffic are to be exercised on an interface. There is a trade-off between making a single, flexible and capable generator or checker, and having several, fixed-function simple classes. Choosing which to use is a judgment call for your team.

These three layers are portable code and can be used for almost any chip. Of course, the generators will have to be constrained, randomized, and started. Also, the checker will have to be waited on until it has checked all the expected chip responses. These activities are the responsibility of the higher test components, as will be described in later sections.

One interesting property of the common currency of interface classes is that each generator and checker probably has at least one thread of execution. This is because hardware is massively parallel, and can operate multiple interfaces independently. In addition, the rate at which the generation and checking occurs is only indirectly tied to the behavior of the interface wires. For example, a single generated "packet" may require many bytes to be transferred at the interface, or several data bytes may be gathered from the interface before the checker is called.

So this is how an interface of a chip is broken down into classes that are manageable and adaptable. The process of examining each chip interface, and then implementing a set of interacting common-currency classes to handle the generating/checking and driving/monitoring, is now repeated for each interface. If this method of using layers and the underlying protocol is well-defined, then there is a good chance that these classes will be used again in later projects.

The Top-Layer Components

The whiteboard drawing is now pretty much converted to code for the chip interfaces. Because we are using an "ends-in" approach, we will tackle the components at the top before we look at the middle layers.

The *top* layer has standard form, roles, and responsibilities, just as the interface layer did. The following figure shows the top layer with its standard classes.

At the very top is the *verification top*, shown in the following figure. This component builds the other top-level components and sequences the initialization, randomization, execution, and shutdown of the simulation. Most existing verification systems have an equivalent top-layer component. Sometimes this component is placed in the testbench class; it's better, however, to abstract its functionality into an independent function. In this way, the specifics of the current project's testbench are removed from the more generic build, startup, and shutdown sequence. The verification top can then be used on multiple projects.

Top-Layer Verification Components

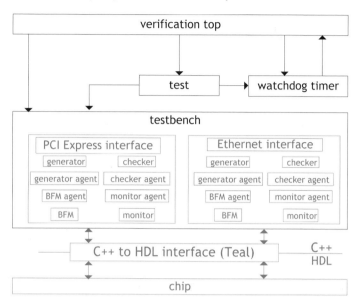

Also at this layer, the three main workhorses of the verification system are the testbench, the test, and the watchdog timer. Like the verification top, the watchdog timer is most likely a generic component. Its role is to shut down the system if too much time has elapsed.

The testbench is probably specific to a project, but it is the same for most tests. Its role is to contain the interface layer objects and perform chip-wide initialization and possibly configuration. The test is responsible for constraining and sequencing the interface-layer objects. (This process is described further in the next section.)

Note that the test changes with each scenario you want to run. The testbench and test implementations differ between projects and runs, yet their interfaces remain constant. At this high level of abstraction, the concepts of building, configuring, running, and shutting down a verification test are uniform.

It's up to the team to decide how to design the interfaces for these standard top-level components, as well as how to design the build/configure/run sequence for a simulation. The Truss chapter has base classes for these standard components.

What is a Test?

The previous section went quickly over the roles and responsibilities of a test. Because a test is an important concept in verification, let's be a bit more thorough. A test is one of the three top-layer classes. It is responsible for exercising some subset of the interfaces of a chip while background traffic or other activity is occurring. The test's main partner is the *testbench*. Before we talk about the test, let's review the role of the testbench and see how the test uses it.

The testbench contains the interface-layer objects for each of the chip's interfaces. In general, the testbench only *holds* these objects; it's up to the test to *use* them. However, there is a small exception: when the chip has mutually exclusive features or interfaces. In this case, the testbench might have an object or two that chooses an appropriate configuration.

The test is responsible for deciding which interfaces or features of the chip are to be tested. Most chips are massively parallel devices. A well-designed test focuses on some part of the chip, but it also exercises other functions or interfaces of the chip simultaneously. Often there is a primary interface or feature to be tested, and a number of independent, secondary interfaces or features.

After a test has "decided" on an interface or feature to focus on, it must constrain the random behavior of those features. The test selects and engages a configuration to the chip, probably by interacting with the interface BFMs. After that, the test starts up all the interface generators and runs them until some end condition is met. This condition could be either elapsed simulation time or whenever the primary interface exercise has completed. Finally, the test waits for all the other interfaces and then reports success or failure.

Because a chip may have several interfaces, it can become tedious and clumsy to work with the interface-layer generators, BFMs, and checkers directly. The test may become cluttered with management code, and it may be difficult (for all but the original writer) to figure out the point of the test. Also, many tests will use many combinations of chip interfaces and features, so that much of the code is replicated. For these reasons, it's better to package a chip's interface-layer test into a concept the authors call a *test component*. Also, the other interfaces that are just exercising the chip as background traffic are packaged into *irritator* components. These components are "middle layers"—that is, they connect the interface layer to the top layer.

Here is an example of what the components of a simple PCI Express test might look like:

Example PCI Express Test

| *Main test part* |
| PCI Express test component |

| *Background traffic* |
| Ethernet irritator |

| *Background traffic* |
| USB irritator |

| *Background traffic* |
| UART irritator |

The test component is described in the next section, and the test irritator is described after that.

To summarize, a testbench holds the interface-layer objects, which the test selects, constrains, and controls. It is good practice to break a test into test components, one for each interface of the chip. For a specific test, a few test components are the main components, while other components—the irritators—provide background traffic.

The Test Component

The whiteboard drawing probably does not include information about the middle layers. This section details some of the questions and decisions related to the middle layer. It is at this middle layer where common questions such as "What object should set what parameters?," "What should be randomized and when?," and "How do we know when we are done?" are answered. In some sense this is the hard part of the verification system, where a lot of mental energy is spent.

The implementation of the middle layer starts with listing the types of exercises you want to perform on an interface or feature of the chip. Often these test requirements take the form of sequences that exercise the basic data paths and functionality of the chip, including error cases.

Once you have this list, you create a middle-layer class to represent each exercise. The authors call these classes the *test components* of a test. A test component does not just represent a stimulus or a scenario for an interface; it also includes the end condition. In a sense, the test component has an interface like that of the verification top, evidence of the fractal nature of a layered approach.

The test component is used to exercise some specific functionality of the chip. In fact, the test component is often used with other test components to create a rich test, with the other test components acting as "background noise" generators (irritators). This is a benefit of designing the test components as a separate class, instead of directly implementing the exercise in the test.

Another benefit of separating the test from the test components is that the test is a layer above the test components, creating them, giving them the appropriate parts of the testbench, and setting their parameters. This allows different tests to drive the same test components differently, perhaps letting a test component "roam" on its parameters, or maybe constraining it to hit a corner case.

In general, a test component, on construction, gets references to a generator, a driver/BFM, and a checker of a chip interface. The references come from the testbench.

Why not take in the entire testbench? By not just referencing an entire testbench and taking the pointers it needs, each test component manages complexity by minimizing the assumptions on the environment. In addition, the test component maximizes the chance that it can be adapted to other testbenches. By having a test component itself perform an exercise, instead of directly implementing the exercise in a test, you have a better chance of ensuring the adaptability of the exercise.

Let's look at a concrete example. Suppose you are driving packets into a chip from one interface and collecting processed packets on another interface. To test this data path, your test system will look something like this:

Test Component Connections

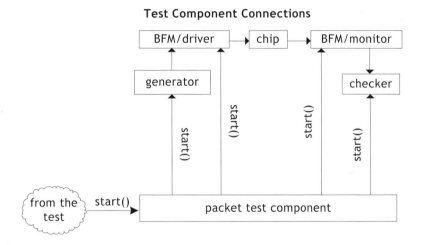

The packet test component class gets a pointer to both BFMs, a generator, and a checker. The role of the packet test component is to exercise some part of the chip by using these other components.

For some methods the test component may just relay calls from the test. In this example, a `start()` method calls `start()` on the generator, checker, driver, and monitor. Recall that the test could have called all the interface layer objects directly; the test component layer just makes the test clearer.

One nonrelay task that the test component performs is to sequence the generator. For example, the packet test component mentioned above would probably tell the generator to generate a certain number of packets.

This number might be set by a randomized parameter, or it could be fixed. There may be other interface-layer parameters that the test component controls, such as packet size or interface configuration parameters. This is where the test component implements what is in the test plan.[4]

The test component is essentially an aggregator. Given pointers to an interface's generator, BFM/driver, and checker, it sequences these as appropriate as the testbench builds the interface components.

The Test Irritator

We have only one more part to consider before we complete the conversion from a whiteboard drawing to a verification system. This last part addresses how to write background traffic components. Recall that a robust test has a test component interface and several other background interfaces. The idea is to ensure that the chip can function in a real-world scenario.

When you start writing tests, you will probably start with test components. Then, after the tests are stable, you'll want to add auxiliary test components. The name the authors use for these auxiliary components is *irritators*. An irritator is most likely to have "evolved" from one of your test components for that chip interface.

When converting a test component to a background traffic generator, you must alter the component so that it addresses not an internally governed amount of traffic but rather an externally controlled one—so that, for example, it does not use a fixed-length group of packets but instead an infinitely repeating sequence of packets. In other words, you want the nonessential irritators to continue doing whatever chip exercise they do until your test says to stop.

The Truss chapter has an `irritator` base class that is inherited from the `test_component` base class. If you write your test irritators so that

---

[4.] A discussion of a test plan is beyond the scope of this handbook, but basically the plan is a list of the exercises you need to perform on the chip.

they use these base classes, irritators can be implemented with very little effort.

By adding irritators, you can write tests that are understandable yet reasonably complex.

A Complete Test

Let's take a step back and look at what we have accomplished. We have progressed from a drawing to classes and code. We have sets of classes for each interface of the chip. We have a testbench that contains instances of each of these classes. We also have a set of test components and irritators that can be combined like building blocks to create a diverse set of tests.

At the top-most layer are the tests. These tests should not only exercise a main function, but also leverage the work of the other team members by using irritators as nonessential traffic. These tests should exercise the chip fairly well.

Shown below is an example of a test following the layers we have talked about, an example UART test with several irritators added.

UART Test Example

This example test includes a "main" test component, called *UART test component*, which probably walks a range of configurations and sends some amount of data. The test also includes Ethernet and PCI Express irritators.

Many choices still remain, such as exactly how to sequence the bring-up, running, and shutdown of these test components. The next part of this handbook provides a standard framework to help you make these choices.

You must still decide how to randomize and constrain the test parameters. For example, there are implementation choices regarding what control variables to include in the test components and the irritators, as opposed to control variables in the generator and configuration for the wire layer. Although these are not simple choices, the next few chapters should help to clarify the trade-offs.

This completes our first pass at converting from a whiteboard drawing to code. As mentioned earlier, this exercise is essentially repeated continually, even as code is written. In other words, reality happens when you write the code.

Summary

This chapter talked about layers. We talked about using layers to increase productivity, by managing the complexity of a verification system.

We talked about "ends-in" coding, where you start at the bottom and top of the test and code towards the middle. We considered this technique of looking at the chip and creating Interface layers as the first step in creating a verification system. We then went to the top layer, and talked about the verification top and the three top components: the watchdog timer, the test, and the testbench.

Next, we entered the middle layer, where we talked about using a test component to exercise a particular configuration or data path of a chip interface. The idea that a test really should have several interfaces exercised at once formed the reasoning behind the irritator layer.

We ended the chapter with a quick tour of a completed test, noting that there are still more decisions to be made as the implementation of the verification system proceeds.

For Further Reading

- *The Mythical Man-Month,* by Fred Brooks, is the classic handbook that talks about why one should not put additional people on a team to solve a problem. He argues, as we did in this chapter, for a more productive team.

- *A Few Good Men from UNIVAC, by* David E. Lundstrom, talks about the concept of a focused and productive team. This book is about the origin of supercomputers.

- *On the Criteria To Be Used in Decomposing Systems into Modules,* by D.L. Parnas, is a 1972 landmark paper on how to go from a problem statement to a design. The fancy name for this process is called *decomposition.* The current fancy terms for thinking about design are "design patterns" or "factory objects." However, be careful with these recent concepts; they refer to good high-level solution templates, but those templates must be applied with care and experience.

- The concept of *design directions* came from Harlan Mills and Niklaus Wirth at IBM in the 1970s. Their original idea was to use a "top-down" approach, but all variants have been popular at various times. The authors believe that an "ends-in" approach is the best for the class of problems encountered in verification.

Part II: An Open-Source Environment with C++

The previous part of the handbook was a high-level look at C++ and how to architect a verification system by using layers. Now we focus on a specific implementation of such a system.

This part of the handbook introduces two open-source libraries, called Teal and Truss, that together implement a verification environment that uses C++. The authors and others have used these libraries at several companies to verify real projects.

The libraries are free and open source because the authors feel strongly that this is the only way to unite and move the industry forward. Locking up people's "infrastructure" is not the way to encourage innovation and standardization—both of which are needed if the verification industry is to improve.

Consequently, you'll find no simulator-company bias in these libraries. These libraries work on all major simulators.

In this part we discuss the following:

- Teal, a C++-to-HDL interface that enables C++ for verification
- Truss, a layered verification framework that defines roles and responsibilities

- How to use Teal and Truss to build a verification system

- A first example, showing how all the parts we talk about fit together

Teal Basics

Coming together is a beginning. Keeping together is progress. Working together is success.

Henry Ford

Building a verification system is a daunting task, but build we must. That is why we use the technique of layering, to break the problem down. By starting with the lowest layer—that is, the one that directly drives and senses the wires—we can start to get some real work done. Still, because C++ is not what most hardware engineers use for their HDL, we'll need an interface layer to connect the HDL with C++. Teal is just such an interface. Teal tries to be as unobtrusive as possible, using terms borrowed from the HDL domain, such as *posedge* and *reg*.

This chapter introduces Teal and shows how to use it. We'll talk a bit about the main parts of Teal—for example, how you can (fairly seamlessly) get and set values in the HDL, and how you can pause execution until HDL signals change.

Overview

• •

Teal is a C++ class library for functional verification. Teal is tiny, consisting of only a handful of source files, yet it provides the necessary minimum features for verification. (A version of Teal is on the companion CD.)

Teal, like Vera, SystemVerilog, and "e," provides the illusion that the verification system is in control of the chip. In Teal, you write a `verification_top()` function, and create tests, generators, checkers, drivers, and monitors. Each of these objects can appear to be running independently of the chip, with each in its own thread of execution. Of course, in reality these threads only execute in response to a chip wire or register change. However, by driving wires and registers, the threads do, in some sense, control the chip.

Teal is unobtrusive; it does not get in the way of your C or C++ structure. You don't put Teal calls everywhere you want to sample or drive a signal, so Teal is also unobtrusive in the HDL code.

The authors realize that many companies have developed their own version of an HDL-to-C/C++ interconnect. We encourage those companies to contact us and share their experiences, so Teal can be made better. This is one of the reasons why Teal is open source.

What Teal provides

Teal enables functional verification by providing connections to HDL signals and allowing actions based on changes in the HDL simulation. It encourages the development of independent generators, checkers, drivers, and monitors by providing management for user-created threads that execute concurrently with the HDL simulation. Teal provides repeatability and constrained random-number generation, as well as a simple interface to pass in runtime arguments, either through the code or command line, or through "scenario" text files. Teal also provides flexible message printing.

This functionality provides the basis for functional verification, but it serves as only a small part of a verification project. You must still write code that stimulates the design, checks the output, and controls the randomness. That is the real work of a verification project. (The next chapter talks about an open-source verification environment.)

Teal's similarity to HDLs

Although Teal does make use of classes and inheritance in C++, your algorithms for driving and sensing the wires can look close to what a hardware engineer is used to.

As an example, suppose you had a signal, located at `top.chip.address`, in your simulation and you wanted to get the value of it at the positive edge of a clock. The Teal code would look this:

```
teal::vreg clock("top.clock");
teal::vreg address("top.chip.address");
teal::vout log("logger id");
at(posedge(clock));
log << "The current address is "<< address" << endm;
```

Don't worry if this example is not clear. We'll walk through each of these Teal classes later. The point is that the `at(posedge(clock))` should be recognizable to Verilog coders. In addition, the address variable can be used as a regular C++ integral variable.

A tiny but complete example

This chapter delves into the details about all of Teal's classes, but let's look at a basic example of what a complete C++ example using Teal looks like. It should not be hard to understand the code presented here, assuming the reader has some familiarity with C++ (or C) and a general knowledge of Verilog (or VHDL).

In this example our chip implements a black-box function. Given a reference clock, it samples a stimulus on the positive edge of the clock and generates a response on the negative edge. To make things interesting, let's assume there is a three-clock latency from the stimulus to the response.

Here's how Teal might be used to drive the stimulus and get the response:

```
#include "teal.h"
using namespace teal;
#include <deque>


void verification_top(){
  const int latency = 3;
  const int number_of_iterations = 10;

  teal::vreg system_clock("testbench.reference_clock");
  std::deque<integer> stimulus_sent;
  vreg stimulus("testbench.stimulus");
  vreg response("testbench.response");

  for (int i = 1; i <= number_of_iterations; i++) {
    //drive the stimulus to the chip and remember it
    at (posedge(system_clock));
    integer stimulus_int; RAND_UINT32(stimulus_int);
    stimulus = stimulus_int; //drive value to the chip
    //save value sent
    stimulus_sent.push_back(stimulus_int);

    //Read from HDL register "response" and print result
    at (negedge(system_clock));
    if (i >= latency) {
      //Note! 'response' in line below reads from HDL
      cout << "For stimulus " << stimulus_sent.front()
           << " the chip produced " << response << endl;
      stimulus_sent.pop_front();
    }
  }
  //need to collect last responses
  for (int i(0); i < latency; ++i) {
    cout << "For stimulus " << stimulus_sent.front()
         << " the chip produced " << response << endl;
    stimulus_sent.pop_front();
  }
}
```

It should be noted that the above example puts all code in the verification_top() function. However, this is not recommended for

• • • • • • •   Hardware Verification with C++

real projects, where a lot more structure is needed (as will be shown throughout this handbook). The point here is that if you use Teal, you won't end up with code that is hard to understand. Teal is straightforward.

In this example we randomized a stimulus input and applied it to the chip, then just printed the response. In a more realistic test, you would have a model of the chip and compare the results to that model.

Teal's Main Components

It is important to decide on a "common currency" when designing a class library. The rest of this chapter describes the common currency of the Teal system—that is, the fundamental building blocks of Teal-based verification.[1]

The following is a summary of the most important classes and namespaces of Teal; more detail is given in the following sections.

- *The* reg *class*—This is one of the most basic classes in Teal. Its main purpose is to provide arbitrary-length, four-state (1, 0, X, Z) "registers" with corresponding operations. The reg class is useful for performing algorithms in the precision of the hardware. It also provides register-slicing operations.

- *The* vreg *class*—This is probably the most commonly used class in Teal, as it connects C++ code to the HDL. The vreg class provides mechanisms to connect wires and registers in the HDL simulation so they can be used in C++ code as though they are built-in C++ variables. The vreg class is inherited from the reg class.

- *The* vout *class*—This Teal class is used for logging, to help trace what happens during a simulation. Modeled after the standard C++ cout object, the vout class provides the ability to report, for example, debug, error, and other informative messages in a consistent format that is coordinated with HDL outputs.

- *The* vlog *class*—This class is a global resource that coordinates all the logging from your C++ code. It receives all vout messages

---

[1.] This is not a complete reference manual, but rather an overview of the capabilities of Teal.

from the simulation and implements a filter chain, so you can add useful features such as replicating output to a file and removing messages or parts of messages.

- *The* memory *namespace*—This namespace provides an abstract interface for reading and writing memory. Internally, a group of memory banks are used to handle memory read and write requests, providing great flexibility.

- *The* vrandom *class*—Because using random numbers for test values is a staple of modern verification, this class is Teal's stable random-number generator. Though small, it provides thread-aware, independent streams of stable random numbers that can be guided by a single master seed. Of course, the numbers all have their own seed as well.

- *The* dictionary *namespace*—This namespace is a global service that abstracts how to set parameters in your test. It provides functionality to get and retrieve parameters from code, the command line, or external "scenario" files.

- *The* run_thread() *function*—This function forks off a thread. You'll use this whenever you have a function, such as a generator or monitor, that needs to operate independently of the test. This function provides a base capability for building transactors, drivers, checkers, and so on.

- *The* at() *function*—This function allows a thread to pause until any of the HDL signals has changed. You provide a sensitivity list of vreg objects, with modifiers such as posedge, negedge, or change. Used with the run_thread() function, it allows several independent tasks to run simultaneously.

All of these classes are described in the following sections, along with the small requirements that Teal puts on the HDL testbench (Teal needs to be initialized from the HDL testbench), and a discussion of how to create the "user-code entry point" function called verification_top().

Using Teal

· ·

It's time to dive into some details regarding how Teal can be used for functional verification. This walk-through of Teal makes it easier to understand the "real world" examples presented in subsequent chapters, while illustrating how Teal can be used in your environment.

Initialization

Let's start at the beginning. For Teal to be used, it must be initialized from the HDL. This is done through an HDL function call that launches Teal. When Teal starts up, it initializes itself, then calls a user-provided function called `verification_top()`.

Verilog is the HDL of choice in this handbook. Because Teal was developed to work with Verilog, many of Teal's syntax and naming conventions mimic those of Verilog.[2]

Teal uses the Programming Language Interface (PLI 1.0 or 2.0) to connect C++ code to HDL simulators. To this end, you must put a PLI call somewhere in the HDL code to start Teal. This call is normally put in an initial block at the top-level testbench, but it can be put anywhere and called at any simulation time. The call is called `$teal_top;` and other than a call for "back-door" memory access, it's the only required HDL call for Teal.

Your Verilog testbench should include the following:

```
module testbench;
  ...
  initial
    $teal_top;
  ...
endmodule;
```

That's all there is to it. Teal will now start and run your test.

[2.] Unfortunately, while a Teal for VHDL is in the works, it wasn't finished in time for publication. Contact the authors at www.trusster.com if you are interested.

Your C++ test

When the simulation begins, the call to $teal_top causes Teal to start the threading system and thereafter call a user-defined function called verification_top(). The verification_top() could be as simple as the following:

```
#include "teal.h"
void verification_top()
{
  teal::vout log("first code");
  log_ << "Hello Verification World" << teal::endm;
}
```

The verification_top() must be defined by the user, or Teal won't link. It normally instantiates other classes and calls their methods. (Subsequent chapters will show example of this.)

Registers

Teal's reg class implements a four-state logic, as well as all commonly supported HDL operations while making sure that X's and Z's propagate correctly. The class supports addition, subtraction, shifting, boolean operations, and comparisons. As in any HDL, bit fields (or subranges) are supported, and they can be on either side of the equal sign. (This will be described in more detail in the next section.)

The vreg class builds on the reg class and adds the connection to an HDL simulation. All events that happen to a wire/reg in either the C++ or HDL simulation get reflected on both sides. The vreg is one of the most used classes in Teal, as it serves as the connection point between HDL and C++.

Creating registers

Creating either a reg or vreg is easy. However, here is one of few places where the two classes differ. When you create a vreg, you supply the string HDL path to the corresponding HDL register, port, or wire you want the variable to reflect. Teal then automatically links together the C++ variable and the HDL signal, and also figures out the correct bit length for the C++ class.

When you create a reg object you don't supply a Verilog path, as there is no connection to an HDL. There are several ways to create a reg. The default is just like in Verilog, a one-bit variable. You can also give the reg an initial value, in which case the register is 64 bits wide. If you need a specific bit width, you can specify that after the initial value.

Here is an example of how you would construct a vreg and a reg:

```
vreg chip_register("testbench.chip.data");
reg cpp_register(0x7FFFFFFFFFFF,47);
```

The first line connects the variable chip_register to the HDL signal located inside the chip instance of the testbench. The second line creates a 47-bit register array and assigns it an initial value.

Working with a reg or vreg

Teal registers are written to act like built-in types as much as possible. This makes working with them easy, and they support assignment to and from most other built-in types; for example, assigning the value of an int to a vreg and vice versa would look like this:

```
int drive_value(0x52571);
vreg v_signal("testbench.chip.signal");
// Assign a value to HDL signal
v_signal = drive_value;
//...
// Sample an HDL signal, assign it to a_sampled_value
int a_sampled_value = v_signal.to_int();
```

The assignment above would work for reg's as well.

In functional verification it is common to access individual fields in a register, whether the contents are individual bits or strings of bits. Teal provides the capability to access both, as the following example shows:

```
uint32 x; RAND32(x); //Assign a random value to x
reg a_reg(x,32);
reg a_field = a_reg(32,25); //bit 32..25 to a_field
cout << a_field;
```

Registers have a number of logical and mathematical functions as well. These functions help define the correct four-state behavior for operations, and make the registers similar to C++'s built-in types.

The following is an example of some of the supported register operations:

```
vreg addr(path + ".address");
addr += 2;
addr = addr >> 2;
addr = addr << 4;
if (addr > 0x64)...
if (addr != 0x1)...
```

Teal does something a little differently when comparing two registers. Because reg is a four-state variable, Teal implements the operator==() as the Verilog triple equal in HDLs. That is, Teal looks at both the 0/1 value and the X/Z value when comparing two registers.

Logging Output

Because a lot of debugging is done by reading simulation log files, in order to see a progression it is important to organize simulations well. In other words, to enable postprocessing, error counting, messages, and possibly filtering, it is important to have a consistent message format. Fortunately, the logging facility in the Teal classes encourages such uniformity. Teal comes with a standardized, customizable logging mechanism, called vout, which mimics C++'s standard streaming mechanism.

Teal uses a two-level logging scheme, as shown in the following figure. In any code that needs to print information, a vout object is created. As many vout objects as needed can be created—which is good, because each vout object can have a relevant instance name.

Each vout object "under the covers" calls a global service vlog object. This is done so that there is a single point of control where reordering, demotion, changing, or deletion of parts of any message can be done.[3]

[3.] Although describing this capability completely is beyond the scope of this handbook, subsequent chapters show several examples.

Vout and Vlog Objects in Teal

Because vout is modeled after the standard C++ library stream std::cout object, vout directly supports the output of the standard types. However, by following a few simple guidelines, you can print complex objects as conveniently as the standard types.

To end a message, just call the Teal endm function. To describe a multiline message, just use endl (like cout) where needed and use a final endm at the end.

When you create a vout, you give it a string that represents the functional area it is in. You can then build any number of message statements. For example, note the following:

```
#include "teal.h"
using namespace teal;
void verification_top() {
  vout log("a test");
  log << teal_info << "val" << hex << 207218 << endm;
}
```

This example prints the following (assuming a thread of tx, a file of uart.cpp, a line number of 313, and a simulation time of 77 ns):

```
[77 ns] [tx] [a test] [uart.cpp] [313] val 64'h32972
```

Teal displays the file and line number in the source code that originated the message. This is useful when the same message comes from several different places in the code. (Of course, this information can be suppressed.)

Note that when you finish a message statement (by using `endm`), the vout instance adds the simulation time, the current thread name, and the functional area to the message, then sends the message to the `vlog` global service. It does not send the message as a text string, which would not allow the efficient modification of the message; rather, it sends the message as a set of pairs of IDs and strings. This allows you to instruct the `vlog` instance to modify messages with respect to their components—for example, to demote errors to a warning, or stop all output from a file or a functional area.

The `vout` class also supports decimal, hexadecimal, and binary output. You select the type of output by placing either a `hex` or `dec` or `bin` in the message statement. The `reg` class also looks at the setting when the `reg` is converted to a string.

However, you often do not need to use the global filtering mechanisms of `vlog`. Instead, you can turn off the display of parts of a message directly, at the `vout` instance. This is described in the Teal reference manual (available on the accompanying CD).

Most verification systems have several levels, or types, of messages. Teal, being no exception, uses the following general categories:

- `teal_info`— Used for standard messages.

- `teal_debug(<level>`—Used when a test wants to display a little more diagnostic information. This is a level-sensitive output; the `vout` class has level-setting methods and accepts a level for debug messages. The message is displayed only if the level of the message is less than or equal to the level that is set.

- `teal_error`—The error type is used when the chip's expected behavior is different from the expected.

- `teal_fatal`—This more-severe error type ends the simulation after displaying the message.

Examples of the above are provided in later examples.

Using Test Parameters

It is often important in functional verification to provide test parameters. These are frequently used, among other things, as constraints for random tests. For, example, a single test case may have several different sets of constraints, each of which covers a selected range of parameters or directs the test into interesting corner cases.

Because such parameters are commonly used, Teal provides a standard, flexible way of working with them. Test parameters can be defined by means of text files, code, or command line entries. Teal handles simple integer and string parameters as well as complex parameters.

The functionality of Teal's parameters is defined in the `dictionary` namespace. Teal maintains a list of parameter names and values, so that a test, for example, can query the dictionary and recover the value.

When you call the `dictionary::read(std::string)` function, Teal reads a text file, takes the first word on each line as the parameter name, and saves the rest of the line as data for that parameter. A special keyword, `#include`, is used to open other files from within files. If a parameter is repeated, the last definition is saved.

In addition to using files, you can also use code to add parameters. When you do this, you have the option of replacing an entry or not.

Parameters can also be entered on the command line. In this case, they override any parameter set by a file or the code. In this way, a parameter can have a default value but still be overridden by a script.

As an example, let's suppose we are testing a UART interface. We have a default parameter file that sets up default constraints, and then each specific test overrides a few values as well as defines its own parameters. The default parameter file could look like this:

```
//in default_parameters.txt:
force_parity_error 0
dma_enable 1
baud_rate 115200 921600
```

A specific parity-error test case could use the default parameter file and override the `force_parity_error` setting like this:

```
//in parity_error_test_parameters.txt:
stop_error_probability_range 32.81962 75.330
#include default_test_parameters.txt
force_parity_error 1
```

The `#include default_test_parameters.txt` line above tells the dictionary to open the `default_test_parameters.txt` file. The `force_parity_error 1` repeats the `force_parity_error` parameter and overrides the default value.

It is not always appropriate to use files to pass parameters. Using files can be good if you need to have many different test parameters and a few basic tests. However, it can be clumsy to make sure the files stay with the respective test code. Therefore, the examples later in this handbook use the code mechanism. Nevertheless, including such files, or even passing parameters on the command line, can be done after most of the test is written, without having to modify the test itself.

So how do we pick up the parameters? The following is a complete basic example of how these parameters could be retrieved:

```
#include "teal.h"
using namspace teal;
void verification_top() {
  // reads file shown above
  dictionary::read ("default_parameters.txt");
  vout log ("first_parameter_example");
  log << teal_info << "force_parity_error is " <<
  dictionary::find ("force_parity_error") << endm;
  }
```

Because most parameters are not strings, Teal provides a templated function, `find()`, to convert parameters to the correct variable format. The `find` function always returns a string—either an empty string ("") if the parameter is not found, or the actual string associated with the parameters. This function relies on the `operator>>` to be defined for the variable class used. The `operator>>` is defined for all built-in types (such as `int`, `char`, `long`, `double`, and so on). For your classes, you can define your own `operator>>` and then use Teal's `find()`.

If defining an `operator>>()` is not appropriate for your class, or if you don't have a class but instead have a collection of built-in types, you can use `std::istringstream`. This allows the code to create a stream from a string, from which you can then extract the chars, ints, doubles, and so on, as needed.

For example, to read the stop_error_probability_range (from the example above), you would use the following:

```
#include "teal.h"
using namespace teal;

void verification_top(){
  dictionary::open("parity_error_test_parameters.txt");
  //reads "32.81962 75.330" from stop_error parameter
      into ss
  std::istringstream ss(
    dictionary::find("stop_error"));
  double stop_error_min (0);
  double stop_error_max (0);
  ss >> stop_error_min >> stop_error_max;
  vout log("showing double double reads");
  log << "Stop error range is "<< stop_error_min <<;
        " to " << stop_error_max << endm;
  }
```

The example above works for all integral built-in types.

Accessing Memory

For most verification projects it is important to be able to access memory. Sometimes you want to do this in zero simulation time. Allowing "back-door" accesses of memory improves simulation performance, allows the monitoring of memory for automatic checking, and makes it possible to insert errors into memory for test purposes. Teal provides such a "back-door" mechanism but also, of course, supports "front-door" access, which can map some memory address ranges to a transactor-based model.

Teal defines each accessible memory (transactor model or memory array) as a memory_bank object. A memory_bank object can be accessed directly through member functions called to_memory() and from_memory(), but each memory can also be associated with an address range, through the add_map() function. In this way, memory can be accessed through addressing by means of read() and write() functions.

Working with address ranges has many advantages, because it creates code that is easier to understand and is closer to production software. The read() and write() functions can even be redefined in production software to become simple integer pointers, as is often appropriate.

When writing a memory transactor, you must define your own memory_bank object, but when working with HDL memories, put a $teal_memory_note() in the HDL. Teal uses this call to make a memory_bank object for you.

The following example shows how HDL memory arrays can be associated with an address range and accessed. An example of how to write memory transactors is in the UART example chapter.

A memory note example

The following diagram shows a small part of a larger testbench structure. This environment verifies a graphics chip that saves graphical texture information in its memory cache. In order to speed up simulation, back-door loading of the texture into the chips memory is used.

Memory Bank Objects

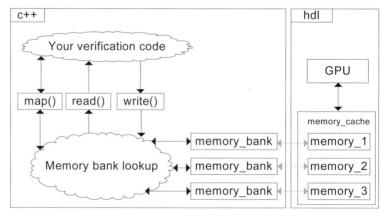

To support direct memory access you need three things. First you need the Teal PLI function $teal_memory_note(), which is called for each memory register array to be accessed. This registers the array with Teal and creates a memory object for that memory. Then you need to deffine an address range for each memory instance to be used, allowing Teal to translate from an address to a specific memory. Finally, once the address range is established, you access that memory through `read()` and `write()` functions.

To do this in the environment pictured above, an initial line block is added to the memory model (that is, the model is, instantiated as `memory1`, `memory2`, and `memory3`, above). So part of the memory model looks like this:

```
reg[31:0] memory_bank_1[1024:0]; //Actual memory array
//to register memory_bank_1 with Teal.
initial $teal_memory_note(memory_bank_1);
reg[31:0] memory_bank_2[1024:0];
initial $teal_memory_note(memory_bank_2);
reg[31:0] memory_bank_3[1024:0];
initial $teal_memory_note(memory_bank_3);
```

As can be seen in the illustration, the memory model gets instantiated three times as `memory1`, `memory2`, and `memory3`. In the `verification_top()` function, the three memories get address ranges declared like this:

```
memory::add_map ("testbench.dut.memory_unit.memory_1",
                 0x100, 0x200);
// The following assumes the subpath memory_2 is unique
memory::add_map ("memory_2", 0x201, 0x400);
memory::add_map ("memory_3", 0x401, 0x600);
```

Now any test can access these memory spaces through simple read and write function calls. Furthermore, neither reading nor writing memory consumes any simulation time. A simple memory access would look like this:

```
memory::write(0x10a, 22); //i.e 0xa in memory_1 = 22
if (memory::read(0x10a) != 22) {
  vlog("memory_example_1") << teal_error
            << "At memory_1[0xa] got "
            << memory::read(0x10a) << " expected 22."
}
```

Note that while the memory is written and read at 0x10A, the actual memory is accessed at 0x0A. This is because of the add_map() that we performed. This allows the rest of the verification system to read and write memory as specified in the chip's memory map. Teal takes care of finding the correct memory bank to access, and then removing the mapping offset.

Constrained Random Numbers

It is important to have a stable, repeatable, seeded random-number generator. Teal's vrandom class provides independent streams of random numbers that can be initialized from a string or file. There are also convenient macros for the most common random calls, such as getting a random integer value or getting a value from within a range.

The rest of this section describes the required initialization of the random generator and some simple examples.

Required initialization

Before using any random numbers you must initialize the random-number generator. This is done by calling the init_with_seed() function and passing it a 64 bit-seed value. It is recommended that higher-level code keep track of this seed value and pass it to the random-number generator. To initialize the random seed generator you would call the following:

```
uint64 master_seed;
...
// master seed gets initilized by higher layer
vrandom::init_with_seed(master_seed);
```

After the random-number generator is initialized, it is ready to be used. The examples in this handbook use the dictionary to get the master seed.

Using random numbers

Because integers are so commonly used here, Teal provides a couple of macros to deal with integers. After you have initialized the random-number generator you can call these macros directly. The most often used macros are RAND_32 and RANDOM_RANGE, which generate a 32-bit random value and a 32-bit value within a range, respectively.

Here are some examples:

```
#include "teal.h"
using namespace teal;
void verification_top() {
  uint32 a_rand32; RAND_32(a_rand32)
  uint32 a_random_range;
        RAND_RANGE(a_random_range, 0, 0x030837);
  vout log (" random number test");
  log << "a_rand32 is " << a_rand32 << endm;
  log << "a_random_range is " << a_random_range << endm;
}
```

When you want to create more-elaborate random numbers, you need to work with the vrandom class directly. The vrandom class is a simple class that you can draw numbers from after it is created. This gives you more direct control over the generation of random numbers. The base vrandom class provides a uniform distribution, but you can create your own classes to have segmented, logarithmic, or other distributions.

You would create an object for your inherited class and draw a number like this:

```
vrandom a_random("some string", some_integer);
uint32 a_random_value = a_random.draw();
```

The macros use the ANSI standard __FILE__ and __LINE__ for the string and the integer. These parameters are hashed with the master seed and are used to initialize this particular random-number generator. You may want to pass in your own values. (For more details, see the reference manual available on the CD.)

Working with Simulation Events and Concurrency

Chips are massively parallel, which means they have many interfaces working at once. So when we test them, we need parallelism in testing as well. Teal provides the ability to start a thread and, if you want, wait for it to complete.

Why does Teal provide this capability when there are already several packages, both operating-system specific and in public domain? Most of the current simulators will "core dump" if any thread runs after control returns to the HDL simulator. Teal's threads ensure that this does not happen. It is this capability that provides the illusion that the C/C++ code is in control of the simulation.

However, having many threads of execution is no good if we cannot pause for some change in an HDL signal or to wait for another verification thread. Fortunately, Teal provides this capability. As soon as you have threads, you'll need a mechanism for exchanging events between threads. Teal calls this mechanism a *semaphore*.

Let's back up a bit and talk about running a thread. The Teal function `run_thread()` allows you to call a c-function in a new thread of execution. This is exactly how Teal starts your `verification_top()` function. The next chapter shows how the `run_thread()` function can be made more "object oriented," but the base mechanism is a c-function. This allows you to decide how object-oriented you want your threads.

It is possible that you may have several `verification_top()` types of functions and want to use that style for starting threads. To wait for a thread to finish, the `thread_join()` function is used. How does Teal know which function to wait on? The `run_thread()` function returns a `thread_id`, which is passed to the `thread_join()` function.

Once you have started a thread, you'll probably set some wires in the chip and then wait for some response. To wait for a wire change, use the `at()` function, which is intended to model the `@(sensitivity list)` statement in Verilog. This function operates on a sensitivity list of `vreg` signals, and the signals are matched on the `posedge`, `negedge`, or any `change`.

Take, for example, a statement such as the following:

```
at(posedge (clk) || change (reset_n));
```

This statement would pause the current thread until either the `clk` signal changed to a one or any change occurred in the `reset_n` signal. Execution would then continue.

As you build layers above this low-level wire layer, the threads themselves need to communicate. Teal's uses its signal and wait class, `semaphore`, to accomplish this. As with threads, the reason Teal provides these capabilities is to prevent a "core dump" in the simulator.

To communicate among threads, two threads need to share a `semaphore` instance. Then one thread (or any number of threads) pauses by means of a `semaphore::wait()` call. Another thread eventually gets some data or reaches some condition and issues a `semaphore::signal()`. That call unblocks the waiting thread. Because you cannot know the order of the thread's execution, a `wait()` may occur after the signal has occurred. The decision regarding whether a thread should honor this previous signal is up to you. If you want to wait for signals that occur only at the current simulation time or later, use the `wait_now()` method.

There is one last point to make about threads. Sometimes you want to make sure that only one thread is using a piece of code at a time. This is common in a BFM that is accessed directly (as opposed to when a queueing mechanism is used). In this case, the BFMs send, read, or write methods must use a `mutex` class. A `mutex` is a mechanism that ensures only one thread uses some shared resource at a time (as will be described in the OOP part of this handbook).

Summary

This chapter introduced an open-source C++-to-HDL interface, called Teal. We talked a bit about how Teal starts up and is connected to your testbench.

Teal's `register` class was covered, along with its inherited class `vreg`. These two common-currency classes are the backbone of the interface to the HDL.

Logging is a very important capability of a verification system. Teal's `vout` class and the global service class `vlog` provide a uniform, yet very flexible, logging capability.

Almost all tests need to have control parameters set by code or files. Teal's dictionary provides a global service for managing parameters.

The `memory` namespace of Teal can be used for both register access and internal chip memory accesses. If reads and writes are extracted from the actual underlying mechanism, different transactors can be used.

Random numbers are essential in verification systems. Teal provides a stable, independent random-number generator.

We ended the chapter with a look at concurrency and Teal's `at()` function. We looked at the `semaphore` class and the `mutex` class for coordinating different threads.

For Further Reading

- The *Teal User's Manual*, available at www.trusster.com, describes Teal in far more depth than this chapter does.

- For connecting the C++ code to the chip, a great handbook is *Principles of Verilog PLI,* by Swapnajit Mitra.

- A standard reference manual on PLI/VPI is *The Verilog PLI Handbook: A Tutorial and Reference Manual on the Verilog Programming Language Interface,* by Stuart Sutherland.

- The pthreads package, officially IEEE POSIX 1003.1c-1995, describes most of the capabilities that a multithreaded system needs.

Truss: A Standard Verification Framework

Truss, and verify.

Anon.

Have you ever watched a building being constructed? Early in the project, when the frame of the building is just a skeleton, it's not clear what the finished building will look like. However, as construction continues, from the windows down to the cubicles that are our workplaces, the intent of the framework becomes clear. In fact, a large part of the building's presence depends on the fundamental structure.

This same basic process occurs when we build a verification system. Early in the project, the application framework is built. The result of years of best practices from both the verification and software fields, Truss is an application framework for verification. It is an *implementation*, and therefore makes some decisions about how things should be structured. With verification as with construction, the framework sets the tone for the system.

Truss is a layered architecture, so you can choose how to implement the layers. Although it makes very minimal assumptions, Truss does provide some base classes and conventions as a guide.

Overview

This chapter presents three main topics:

- The roles and responsibilities of the various major Truss components
- How these components work together
- How you adapt this framework for your verification system

This chapter builds on the two previous chapters of the handbook. It implements an open-source verification infrastructure based on the discussion in the Layered Approach chapter. It also uses the Teal library described in the last chapter as a connection between C++ and the simulation.

Teal provides the fundamental elements of a verification system and supports a wide array of methodologies. Truss, on the other hand, provides the infrastructure layers above Teal, adding a set of classes, templates, idioms, and conventions to facilitate the construction of an adaptable verification system.

One of the tricks in building a reasonable system is to find the *key algorithm*. The rest of the algorithms can usually fit around that key algorithm. For example, in a video editing program the key algorithm is all about getting the pace of the edits right. When you watch a movie, that happy, sad, or scared feeling you get comes from how well-timed and precise the changes in scene are.[1] The authors, having developed software for video editing systems, know that in this domain the key algorithm is implemented by adjusting the edit points of a few seconds of video while the video is constantly looping around those edits. This is not a trivial thing to do, because multiple streams of video and audio,

---

[1] Okay, emotions also come from the music, but everything works together.

possibly with software algorithms to implement effects, are changing as the user is adjusting the edit points.

In the verification domain, the key algorithm is the sequencing of the various components of the system. The authors refer to this as "the dance," as there are usually a few interacting components involved. As we talked about in the Layered Approach chapter, the top-level dance takes place between the test, the testbench, and the watchdog timer. Truss implements this dance in the `verification_top()` function—but Truss does not stop there. The authors believe that this dance is the key algorithm in several layers of the system, so we created a `verification_component` abstract base class. Also, we created `test_component` and `irritator` base classes to be the "top" at the interface and feature layers of the system. Recognizing and reusing the dance is a significant part of Truss.

This chapter explains the major components of Truss, providing code examples where appropriate. Subsequent chapters provide more-detailed examples.

General Considerations

The authors have worked on several different implementations of verification systems before Truss was available. While at a high level verification systems can be described uniformly, the language used to build them has a lot to do with how a specific framework is constructed.

Using a language other than C++

It is possible to build an OOP-based verification framework in languages other than C++, but no other verification language on the market has the OOP capabilities of C++. For example, when a language that does not support operator overloading is used, the generic `operator==()` or copy constructor cannot be used. To provide this basic required functionality, a common generic base must then be used. Unfortunately, this warps the framework and produces a fragile architecture—mostly because of the unsafe type casting. As another example, with a language that has a compilation library (such as current HDLs), there is usually a failure to make a distinction between interface and implementation. This leads to a more-complicated framework, as test writers must separate the interface from the implementation manually and repeatedly. C++ avoids these problems.

Keeping it simple

A stated goal with both Teal and Truss is to avoid unnecessarily complicated code. C++ has many powerful features, but many times they are not appropriate. It is easy to get distracted with C++ techniques and forget that the real goal is keeping the whole team productive.

For example, implementing a generic interface for a verification component, such as a transactor, as a template can be tricky. Sometimes using the template can be more complicated than simply replicating code.

Sometimes only a convention should be used. An example of this is the *generator* concept. One could define an abstract base class, yet the common methods come down to just `start()`, `stop()`, `report()`, and a few others. It turns out that this concept of `start()`, `stop()`, and so on is common to a large set of verification tasks, and is represented in Truss as the abstract base class `verification_component`. However, the concrete subclasses are inherited from `verification_component` only if they use the bulk of the methods. Any smaller subset uses the same named methods as a convention instead.

> *In this way, the framework is not warped to fit a generic class. Even more important, your design is not warped to fit the generic class.*

Truss implements a specific methodology for functional verification. As in any endeavor to generalize, the terrain is fraught with peril. Nevertheless, as writing code entails making judgments about what is the "right" decision, Truss attempts to generalize a style of verification. Deciding on the right balance between generic and specific is a judgment call for the team. The idea behind Truss is to foster a small, usable, and adaptable methodology for beginners through experts. As such, Truss provides an example of the techniques presented in Part III of this handbook.

Major Classes and Their Roles

Truss is an implementation of the layers talked about in the Layered Approach chapter. Consequently, there are only a few top-level components—the verification top, the testbench, the test, and the watchdog timer. Each component has a specific role. These components and their roles have been architected to allow a large amount of flexibility with a relatively simple interface. These top-level components (and those the next level down) are shown below:

Verification Component Hierachy

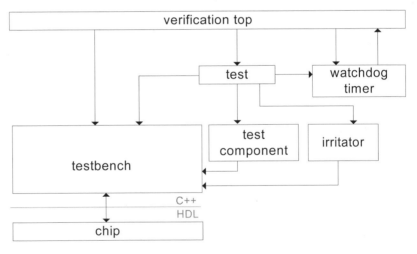

The top-most C++ component is the `verification_top()` function, whose role is to create and sequence the other components through a standard test algorithm. (The algorithm is explained in detail in the next section.) In addition, `verification_top()` initializes all global services, such as logging, randomization, and the dictionary.

The watchdog timer is a component created by `verification_top()`. This component's role is to shut down a simulation after a certain amount of time has elapsed, to make sure the simulation does not run forever.

The testbench top-level component is the bridge between the C++ verification world and the HDL chip world. As such, the testbench's role is to isolate the tests (and test writers!) from having to know how C++ transactors, traffic generators, monitors, and so on interact with the chip. Whether a bus functional model (BFM) writes to registers or forces wires should not be of concern to the test writer.

In addition, the testbench holds the configuration objects of the chip. This is needed by the BFMs, transactors, and similar agents to be able to configure the chip correctly. There is probably a configuration object for each interface of the chip. For chips that contain internal functions, such as dynamic memory allocation (DMA), there may be a configuration object for each function.

The last, but certainly not the least, top-level component is the test itself, whose role is to execute a specific functionality of the chip. It does this by using the testbench-created BFMs, monitors, and generators. The test is responsible for choosing among the testbench's many configurations and capabilities and exercising some subset of the chip's functionality. In general, the test contains very little code. This is because any code it contains may need to be used in other tests as well. To support code that is more adaptable, a test normally consists of several test components, as will be discussed later. The exception is for directed tests, in which case registers may be overwritten, specific traffic patters sent, or specific corner cases exercised directly in the test component.

Key test algorithm: The "dance"

The top-level components of the previous section have a complex, yet necessary, set of interactions. This ensures the maximum flexibility for a test, while providing a known set of interactions. This is one of the tricky parts of a verification system. This section discusses this standard algorithm, which we call the "dance."

In general, the top-level components are created, randomized, and then started. Then `verification_top()` waits for the test and testbench to be completed. This is called the "polite" path. If the watchdog timer decides that a timeout has occurred, the "impolite" path is taken and the simulation ends.

The order of these calls can be better visualized on an event diagram, as shown below. The four columns show the main components. Execution starts at the top left line, and the arrows represent function calls to the other components.

The Dance

The first thing that `verification_top()` does is build the global logging objects. These provide logging to a file and shut down the simulation after a threshold number of errors have been logged. (See `truss_vout.h`.)

Then, after the global logging objects have been created, `verification_top()` allocates the top-level objects. The test is given a pointer to the testbench, so that it can interact with the testbench. It is also given a pointer to the watchdog timer, in case a part of the test wants to force a shutdown or override the `verification_top()` default time-outs. The watchdog is given a pointer to an object so that it can call the final report method with a "watchdog timeout" string prefix.

At this point, all the top-level objects are constructed. As part of their construction they are expected to have established default constraints.

Then `verification_top()` reads the dictionary file (if it exists). This is to allow the test constraints file to override any default settings put there during the construction of the test, the testbench, and their subordinate components.

After initializing the random-number generator, `verification_top()` calls `test->randomize()`. Once the test is randomized, then `testbench->randomize()` is called.

At this point, it is expected that the test and testbench have built their respective subcomponents and are ready to run the test. The first step is the `time_zero_setup()` method, which is used to force wires and initialize interfaces prior to bringing the chip out of reset.

As expected, the next step is `out_of_reset()`, which is used to bring the chip out of its reset state and set it for initialization through the backdoor or register writes.

The next step, `write_to_hardware()`, is where the BFMs are called to initialize the chip. This can be done by either the test, the testbench, or a combination of the two. What is appropriate depends on your situation, as discussed in subsequent sections.

At this point the system is ready for traffic flow. The `start()` method directs the testbench and test to start running. The testbench is started first, to allow monitors and BFMs to start, followed by the watchdog timer. Finally, the test is told to `start()`, which generates the actual traffic.

Next, `verification_top()` calls `wait_for_completion()` on the *testbench*. If your design makes the testbench aware of what checkers are in use, this call waits for the testbench checkers to complete. If not, this method simply returns.

Then `verification_top()` calls the *test*'s `wait_for_completion()`. If your design makes the test aware of what checkers are in use, this call waits for them to complete. (This is the style used in the examples.)

At this point, the test is almost finished. The testbench and test are called to report their final status.[2]

Then `verification_top()` checks to see if any errors were reported. If none were reported, the test is considered to have passed. It may seem weak to accept that the absence of errors is sufficient to consider a test passing. In practice, however, there is no other choice. At the top level, one must trust that the lower-level objects do their jobs. Note that this usually means that in-flight data must be weeded out as the checker proceeds.

Now if the watchdog timer triggers, a different path is taken. The watchdog immediately calls the report method on `verification_top()`. Note that the watchdog itself uses an HDL-based timeout, so that if the report method hangs, the simulation still ends.

The verification_component Abstract Base Class

. . . .

While the test and the testbench are completely different classes as far as their roles and responsibilities are concerned, their interface to `verification_top()` is the same. For this reason a common class was created. This common class, used as a base for both the test and testbench, is called the `verification_component`.

The `verification_component` is an abstract base class. As such, it provides pure virtual methods for the dance described in the previous section. In addition, `verification_component` provides a constant

---

[2.] The authors have tried using the destructor as the final report mechanism. In practice, however, this becomes a difficult part of the design. This is because some destructors try to access deallocated memory or other objects that have already been destroyed. It then becomes tricky to "shut down" the simulation in the correct order, so as not to cause a crash or hang and still get errors printed out. This is one area where verification is different from software, which generally does use destructors as part of the system design.

name and a logger. The interface for `verification_component` is shown below:

```
namespace truss {
  class verification_component {
    public:
      verification_component(const std::string& n);
      virtual ~verification_component();
      virtual void randomize() = 0;
      virtual void time_zero_setup() = 0;
      typedef enum {cold, warm} reset;
      virtual void out_of_reset(reset) = 0;
      virtual void write_to_hardware() = 0;
      virtual void start() = 0;
      virtual void stop() = 0;
      virtual void wait_for_completion() = 0;
      virtual void report(const std::string prefix) = 0;
      const std::string& name;
    protected:
      mutable teal::vout log_;
  };
};
```

Although `verification_component` is a base for the test and the test-bench, it is also useful as a base for other objects.

Detailed Responsibilities of the Major Components

The previous sections discussed the roles of the major components and how they were sequenced to run a test scenario. This section dives down a level, discussing in more detail the specifications of the major components. (Because `verification_top()` was discussed in detail in the previous section, it is not discussed further here.)

The testbench class

The `testbench` class has two main responsibilities. One is to isolate the test writers from the actual wire interfaces. The other is to provide "one-stop shopping" for all the generators, checkers, monitors, configuration objects, and BFMs/drivers in the system. The reason to put all of your components into a single object is to facilitate the adaptation of components into multiple tests. In this way, a test writer can see all of the possible "building blocks" that are available.

The `testbench` class can be a passive collection point for all these components, or it can play an active role in bringing the chip out of reset, generating traffic, and knowing when the test is done. In theory, only the global functionality should be handled by the testbench. For example, the testbench probably should bring the entire chip out of reset, while the test can bring separate functionality out of reset. In practice, the test and the testbench share the work.

In general, it is better to let the test or test components control the simulation. This is because a test or test component can then be adapted for several different types of tests.

A more active testbench may, as a counterpoint, simplify a large number of tests in a way that a test base class cannot, because the testbench has direct access to all the chip's wires.

Understand that the more test knowledge a testbench has, the more all tests must act the same or have control over that testbench's functions. This can be good or bad. The specific responsibilities for control and functionality—test or testbench—are, of course, up to the verification team.

As an implementation detail, Truss provides only a `testbench_base` class. What `verification_top()` builds, however, is a `testbench` object. You must provide a `testbench.h`, which declares a `testbench` class. You will probably also have a `testbench.cpp`, which is inherited from `truss::testbench_base`.

Watchdog timer

The watchdog timer component is responsible for providing an "impolite" shutdown if the test has executed for too long. The timer has two timeout mechanisms: one triggers when the watchdog HDL timer triggers, and the other triggers after the first trigger has occurred.[3]

The watchdog timer uses the dictionary to get its timeout values, which are sent to the HDL on `time_zero_setup()`. The `start()` method starts the timers. The HDL watchdog uses an internal timer. If it were to use a passed-in clock, that clock may inadvertently be shut off.

Once either timer triggers, the watchdog HDL timer is notified and a second timer is started. If this timer expires, `$finish` is called. This might happen, for example, if there is some code in the report that is still reading registers, but the chip is unable to respond.[4]

After the watchdog is notified of an HDL timeout, the `report()` method in `verification_top()` is called. This allows the test to report which checkers have completed and which have not, helping to provide a clue as to why the simulation ran too long.

Test class

The test class is responsible for selecting, configuring, and running all the appropriate generators, BFMs, monitors, and checkers. It is also responsible for selecting the configuration of the chip to be used.

While you could directly implement the above responsibilities in the test class, Truss encourages another style. In Truss the test is intended to consist of a number of independent, smaller components called *test components*. These components are the ones that actually do the work; the test's role is to create, constrain, configure, and sequence the com-

---

[3.] The watchdog timer is simple in theory, but often hard to execute correctly. To be sure, it must have a clock and a countdown time, but even this basic level can be problematic. Should you use wall clock time, simulation time, or both? Should the HDL timer be internal or external? What resolution should it have? Should the test be able to extend or communicate the expected time of the run?

[4.] The authors worked on a project where the final report code read the status registers to make sure that functional area of the chip did not have any errors. However, when we added a power-down test irritator, the read hung the system. It took us a while to find the offending code.

• • • • • • •

ponents, as appropriate for the test at hand. The reasoning behind having multiple independent components is that this is close to the real operation of the chip, where each feature is expected to operate simultaneously. In reality, the chip has common resources that must sequence or arbitrate the use of features. It is in these common resources where the more tricky bugs lurk.

Using this method, the test's direct responsibility is to map the features of the chip (as presented by the testbench's data members) to a set of classes inherited from the `test_component` base class. The test would then add constraints to adapt the test component to the test at hand, as in the following example:

```
class ethernet_basic_packet : public test_base {
  public:
  ethernet_basic_packet(testbench* tb, watchdog* wd) :
    ethernet_data_1(tb->e_generator_1, tb->e_bfm_1,
                    tb->e_checker_1),
    ethernet_data_2(tb->e_generator_2, tb->e_bfm_2,
                    tb->e_checker_2),
    pci_express_1(tb->pci_generator_1, tb->pci_bfm_1,
                  tb->pci_checker_1) {}
  void time_zero_setup(){
    ethernet_data_1.time_zero_setup();
    ethernet_data_2.time_zero_setup();
    pci_express_1.time_zero_setup();
  }
  void out_of_reset(reset r) {
    ethernet_data_1.out_of_reset(r);
    ethernet_data_2.out_of_reset(r);
    pci_express_1.out_of_reset(r);
  }
  void write_to_hardware() {
    ethernet_data_1.write_to_hardware();
    ethernet_data_2.write_to_hardware();
    pci_express_1.write_to_hardware();
  }
   void start(){
    ethernet_data_1.start();
    ethernet_data_2.start();
    pci_express_1.start();
  }
```

```
    void wait_for_completion() {
     ethernet_data_1.wait_for_completion();
     ethernet_data_2.wait_for_completion();
     pci_express_1.wait_for_completion();
    }
    void report(const std::string& prefix) {
     ethernet_data_1.report(prefix);
     ethernet_data_2.report(prefix);
     pci_express_1.report(prefix);
    }
  private:
    ethernet_test_component ethernet_data_1;
    ethernet_test_component ethernet_data_2;
    pci_irritator pci_express_1;
  }
```

In the above example, the `ethernet_basic_packet` test uses three test components, two of which are identical. It connects up the appropriate testbench objects and forwards to every test component the following test calls:

`time_zero_setup()`, `out_of_reset()`, `start()`, `wait_for_completion()`, and `report()`

So why do testing in this more complicated manner? In addition to the previously mentioned idea of simulating close to real-world conditions, an important reason is to maximize the adaptability of the test components. In the example above, we used the same test component for both Ethernet ports. Also, when the test components take in only the parts of the testbench that they need, they (1) make explicit what they are using, and (2) minimize the assumptions on the rest of the chip. This, as will be highlighted in the single UART example in Part IV, allows a test component to be reused for other chips that have only a subset of the original chip's functionality.

Test components are critical to the adaptability of a verification system. In general, the test components themselves do not know whether they are running in parallel with other test components or are part of a series. Thus, the most adaptable components are these test components, as will be discussed further in the following sections.

As an implementation trick, `verification_top()` builds a test by using a `define` called TEST. This trickery, set up by the makefile, allows the `truss` run script to compile in a different test, while leaving the rest of the build image the same for all tests. This allows each test to be its own class (inherited from `test_base`). This cleverness helps one avoid a bad experience in the future. Assume that your team had written on the order of 50 tests, and then a new test was created that required a new subphase to be added to the dance. Although the other tests did not need this new method, you cannot add the default method. This is because all the tests are implemented as a test class. There is only one header `test.h`, and 50 different `test.cpp` files. By defining a base class, and then having the actual test be an inherited class (with a different header file), one can add methods to the base without affecting the existing tests.

There is one more part to a test that needs to be discussed. Often a test is made better by the addition of random background traffic. This traffic, be it register reads and writes, memory accesses, or just the use of other interfaces, can uncover corner cases, such as bus contention, that would not be found otherwise.

These background-traffic test components are called *irritators* and inherit from the `test_component` class. They differ from the standard test component in that they continue their traffic generation until told to stop by the test. Test components, by contrast, decide themselves when they are done, as determined by specified metrics, such as a stop time or the number of packets to send. (Irritators will be describe in more detail later in this chapter.)

With background traffic irritators, the test is written essentially as before. The exception is that the `wait_for_completion()` of the test calls the primary test components' `wait_for_completion()`. When the primary component returns, the test calls `stop_generation()` on all the irritators and waits for them by means of their `wait_for_completion()`. Then the test returns control to `user_main`. (This is explained further in subsequent sections and in the examples in the chapters that follow.)

Test Component and Irritator Classes

As discussed in the previous section, test component-based design is central to a Truss-based test system. The authors have found that separating the test scenarios into test components has maximized the adaptability of the system. By using test components and irritators, test writers have been able to minimize their assumptions and distractions and concentrate on exercising the chip. Furthermore, other test writers can adapt what was done in other functional areas and inherit irritators (if they are not already present) for use as background traffic.

This section describes the responsibilities and interfaces of the `test_component` and `irritator` abstract base classes.

The test component abstract base class

The `test_component` is an abstract base class whose role is to exercise some interface of the chip. As discussed above, this functionality has traditionally been included in the test. The `test_component` describes the interface that all concrete implementations must follow.

In fact, you may have several types of `test_component` for a single interface, for example, a register read/write one, a basic data path one, and an error case one. The fact that these different exercises implement the same interface simplifies reasoning about them.

In practice, most test components use a generator and a wire-level object. Sometimes they may also be given a checker, depending on the designer's intent.

The `test_component` class is not directly a `verification_component`, but it has all the same phases.[5] The `test_component` breaks down some

---

[5.] The primary reason for this is because `verification_component` represents a pattern, while `test_component` is an example of this pattern. The `test_component` has specific implementations of four of the `verification_component` methods. Also, `test_component` introduces some of these same methods as nonvirtual. Finally, the sequencing of the methods is different from the test and testbench, the two top-level components that are verification components. These differences are critical for the integrity of the class.

of the `verification_component` methods into finer detail, as one would expect of a lower-level object.

Below is the interface for the `test_component` base class.

```
namespace truss {
  class test_component :
      protected virtual verification_component,
      protected thread {
    public:
      test_component(const std::string& n);
      virtual void time_zero_setup() = 0;
      virtual void out_of_reset(reset) = 0;
      virtual void randomize() = 0;
      virtual void write_to_hardware() = 0;
      void start();
      void stop();
      void wait_for_completion();
      void report(const std::string& prefix);
    protected:
      virtual void start_();
      virtual void run_component_traffic_();
      virtual void start_components_() = 0;
      virtual void generate() = 0;
      virtual void wait_for_completion_() = 0;
      bool completed_;
  };
}
```

The methods `time_zero_setup()`, `out_of_reset()`, and `write_to_hardware()` are provided to allow the test component to interact with a BFM or driver. Note that a different, but equally valid, architecture would keep the wire-layer components private in the test-bench and sequence them by means of the top-level dance. This assumes that the testbench knows what subset of the BFMs, drivers, and monitors, to start up.

The `start()` method is used to start the `test_component`'s generator, BFM, and so on. This method is implemented by a Truss utility class called `thread`. A `thread` class runs another virtual method, `start_()`, in a separate thread or execution. This allows a test class to do the obvious thing and just call `start()` on all the test components the test uses.

Let's look at the `start_()` method, as it is the main starting point for an interface of the chip. The `start_()` method runs two methods: a `start_components()` pure virtual method, and a virtual `run_component_traffic_()` with a default implementation. The idea behind the `start_components_()` method is that you call `start()` on your generators, BFMs, and so on, as appropriate. (The examples part of this handbook contains examples of `test_component`.)

The default `run_component_traffic_()` method calls `randomize()` (to randomize the test component and its components), and then calls `generate()`. In your `randomize()` method, randomize the data members that will be used by `generate()` to cause some traffic to be generated. In your `generate()`, take these data members and make the appropriate calls to the generators in the testbench.

An AHB example

An example might make the roles a little clearer. (Remember that there are several fully implemented examples in Part IV.) Suppose you are creating a test component to test an AHB[6] arbiter. The test component acts as a master, generating read and write requests to a number of slaves.

The generator in the testbench can generate a burst of reads or writes to a given slave, using a specific burst length. Assume that the generator has a channel interface that can take in an AHB transaction object. The randomize function of your `ahb_test_component` might look like this:

```
void ahb_test_component::randomize() {
  burst_length_ = generate_burst_length(min,max);
  is_read_ = generate_type(min_type, max_type);
  slave_ = generate_slave(min_slave, max_slave);
}
```

The corresponding `generate()` might look like this:

```
void ahm_test_component::generate() {
  //addresses are picked by the generator
  generator_->queue_burst
    (new AHB_transaction (burst_length, is_read_,
```

6. AMBA (Advanced Microcontroller Bus Architecture) high-performance bus.

```
        slave_ ));
  done_.signal();  //Signals that test_component is done
  }
```

Notice that by nature these calls are executed in a one-shot manner. That is, together they perform a single transaction. This is useful to allow an `irritator` to inherit from this test component later, to sequence this pattern any number of times and possibly change the randomization constraints as well.

So why have two separate methods?

By separating the randomization from the generation phases, one can inherit different classes that either (1) have different randomization characteristics (for example, logarithmic distributions of the burst length, or a pattern); or (2) send the data through a filter first, then to the generator.

So now that the transaction has been generated, what should the `wait_for_completion()` method do? Because the generation is occurring in another thread, there should be a condition variable to communicate when it is done.

So the code might look like this:

```
void AHB_test_component::wait_for_completion_ () {
  done_.wait();
  }
```

Test-component housekeeping functionality

The `test_component` class also provides a basic housekeeping boolean that tracks when you return from the `wait_for_completion_()` method. This allows the `report()` method to determine whether you have considered the work of the component to have been completed or not. This can be very useful in a timeout situation, to see which components have not completed.

What you decide to do in the `wait_for_completion_()` depends on how you view your `test_component`. One view is that it is a traffic generator only, which can complete when the generation of traffic has been queued. It is then up to the testbench or test to determine when the chip has processed all the data. This will most likely involve a checker or monitor.

Another view is that your `test_component` represents a generate and check path through the chip. In this case, the completion of `test_component` signifies the completion of the entire exercise. (The examples in this handbook use this view.)

As always, the team must decide which view is better for their project.

The irritator abstract base class

As discussed above, the `test_component` is set up as a one-shot traffic generator. This works for tests that are directed, and for tests where the completion event is predetermined—that is, tests that know before the `start()` call what the end conditions are.

However, sometimes it is not good design to have the `test_component` determine when completion is achieved. This is the case when, for example, you want to achieve a certain metric, and the measurement is not appropriate information for the `test_component`.

For example, you may want to send 100 bursts of some AHB traffic. While this could be included in the `ahb_test_component`, you might not want to measure completion by 100 bursts all the time. Instead, you might want to write a test that looks at the number of hits each slave device gets, and stop the test when all slave devices have been targeted. As another alternative, you might want a test to run until some coverage occurs, which could be any of the previous scenarios, or could involve some internal state in the arbiter.

The `irritator`, inherited from `test_component`, is used for situations such as these. The interface is shown below.

```
namespace truss {
  class irritator : public virtual test_component {
  public:
      irritator(const std::string& n);
      virtual ~irritator() {}
      void stop_generation() {generate_ = false;}
  protected:
    virtual void start_();
    virtual void run_traffic_();
```

```
        virtual bool continue_generation();
        virtual void inter_generate_gap() = 0;
        bool generate_;
    };
  };
```

The irritator overrides the `run_traffic_()` method of the `test_component` base class. It sets up a loop, calling the one-shot randomization and generation in the `test_component`'s `run_traffic_()` methods. The implementation is shown below.

```
  virtual void truss::irritator::run_traffic_() {
    while (continue_generation()) {
      test_component::run_component_traffic_();
      intergenerate_gap();
    }
  }
```

The method `continue_generation()` just looks at a boolean, which is toggled to `false` by a call to the `stop_generation()` method. This allows an external class to stop the continual loop of randomization and generation.

Note that there is a new virtual method in the `irritator` class, called `intergenerate_gap()`. Because the irritator is continually generating traffic, you might need a delay mechanism to prevent the generator from flooding the chip.

There are many ways to get this delay. For example, in one solution the generator and attached BFM/driver could execute the generate request as soon as it is called and thus take simulation time. In another solution, the way to get a delay would be to have a fixed-depth generator and BFM/driver channel.[7] This would put back-pressure on this generate loop. In still another solution, the generator could have a delay in clock cycles before returning.

Any of the above solutions is acceptable, but there is yet another choice. That option is to have the irritator itself provide the delay mechanism.

[7.] This method is supported in Truss's `channel` class.

The `intergenerate_gap()` is a virtual method allowing you to implement an irritator-based delay. This allows the irritator to decide on the throttle mechanism. Different subclasses could implement different policies. For example, an irritator could wait for a variable number of clock cycles. Another example would be to measure some parameter on the checker (such as packets in flight).

As always, the team must decide what is appropriate.

Using the irritator

The irritator continues this generate/wait loop until a `stop_generation()` is called. But how do you decide when to stop the irritator? The answer, of course, is "When the test reaches its goal." One goal could be that the "main reason" for the test has been achieved. For example, you can have the main goal be a test component, perhaps one that generates a fixed, but randomized, number of packets through a particular chip interface. The global goal in this case would be for the test component to achieve completion. Here is how the test code might look:

```
void noisy_packet_test::wait_for_completion() {
  //assume the data members include
    base_packet_exerciser,
  //the test component of interest and some std
    container
  //class with a list of irritators.
  basic_packet_exerciser_->wait_for_completion();
  std::for_each(irritators_.begin(), irritators_.end(),
            stop_generation());
  std::for_each(irritators_.begin(), irritators_.end(),
            wait_for_competion());
}
```

Ignoring the nontrivial constraining, selecting, and creating of the test component and irritators, what is accomplished in a few lines of code is a shutdown sequence that is powerful, while being a fairly simple idiom.

Note that a verification team could decide to use only irritators in their implementation. In that way, when to stop the test can then be determined by looking either at a checker or possibly at elapsed simulation time.

The complex part of the test would then become the randomization and selection of irritators. The authors have worked on a variant of this methodology, and the resulting verified chip was a first silicon success.

Summary

This chapter introduced Truss, an open-source application framework.

We revisited the benefits of an OOP language such as C++, but stressed the need to keep things simple despite the power of this language, to avoid writing code that is unnecessarily complicated.

We talked about the key algorithm of verification, which the authors called the "dance." We showed how the dance is used by the `verification_top()` program to run a test. We discussed the roles and responsibilities of the test, testbench, and watchdog timer, the main parts of the top-level dance.

We discussed the `verification_component` abstract base class, which provides pure virtual methods for the dance.

We then discussed the `test_component` and `irritator` classes, including their responsibilities and interfaces.

Truss Flow

*Expensive solutions to all kinds of problems
are often signs of mediocrity.*

Ingvar Kamprad, founder of IKEA

Have you ever bought and assembled a piece of furniture from IKEA?
In the store most of their furniture looks very simple, but when you get
it home and try to assemble it, you realize that it's built from several
smaller and often confusing pieces. Even with IKEA's famous assembly
instructions, showing the "intent" for each piece graphically, assembly
can still be confusing. Imagine how hard it would be without instructions.

The authors have had to learn many verification environments through
the years, and this has often been a very confusing experience. What
seems like a great concept with a well-defined structure at a high level
of abstraction is often obscured by troublesome details when you first
try to implement it. Many times the confusion is increased because of a
lack of description regarding how the high-level ideas are actually implemented. To help reduce the confusion around Truss, this chapter describes
the *"dance"* in more detail.

Overview

This chapter looks at how the "dance" described in the preceding chapter is actually implemented. It shows the order in which each method is called, and describes the files to find the method, or its base. The chapter then looks at the structure for the major components of Truss.

First to be described is `verification_top()`, the first function called in Truss and the base of the "dance." Following this is a description of the methods, and their class, through which files are called for each step.

Then the test component is described. This component follows a dance similar to that of `verification_top()`, but for a different set of classes and files.

The `irritator` class is described next. While similar to a `test_component`, irritators have some unique method calls worth pointing out.

The last part of the chapter talks about steps that need to be taken to build a new Truss project, by taking the more-abstract description of classes and applying them to the first few tests in a new project.

About verification_top.cpp

When the simulator executes the `$teal_top` call in the HDL, control is passed to the `verification_top()` function in `verification_top.cpp` under the *truss* directory. In this handbook we refer to this function as the "dance," or top function. It is this function that interacts with your top-level components: the test, the testbench, and the watchdog timer.

Let's look at the dance with respect to the methods you have to write. This is illustrated in the figure on the following page. A square box indicates that the method has a default implementation, and a rounded box indicates it needs to be defined for your project.

The Dance – Detailed Flow

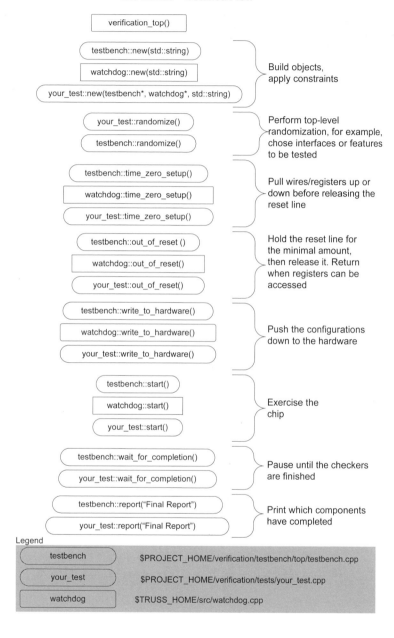

The `watchdog` class is already written and should be sufficient for most purposes. (We will not discuss the watchdog timer's methods, because they are relatively straightforward.) You'll have to write the test and testbench classes.

In the `testbench` constructor, instantiate your generators, checkers, BFMs, and so on. (This assumes that your team has decided to put these interface objects in the testbench rather than in the test components.) Then add your constraints by using the *dictionary*. These constraints will be picked up by your generators and configuration objects to guide the randomization. Initially, you will probably have no constraints.

The test's constructor will create all the test components and irritators that it needs.

In the `testbench::randomize()` method, randomize your local variables and then call `randomize()` on lower-level components, as appropriate. Your testbench may have configuration objects for each interface or feature that is used to configure the chip.

The `test::randomize()` method is similar, in that the test randomizes each `test_component` it owns. In addition, the test may select some subset of all the components and irritators it owns.

The `testbench::time_zero_setup()` method is where you drive wires prior to letting the chip out of reset. You may need to wait for the PLL to lock, or set up "sensor" pins on the chip in this method.

The `test::time_zero_setup()` usually just calls all the active test component's `time_zero_setup()`. This is to allow test components that have a "plug-in" behavior, such as USB and PCI Express, to perform their initial training. (To use this method is a judgment call, as you may want to bring up an interface later in the simulation.)

The `testbench::out_of_reset()` will bring the chip to a stable state that can accept register access. If the team so decides, you could use `test::out_of_reset()` to reset the chip.

The `write_to_hardware()` methods in both the test and the testbench are where you perform register writes to move your selected configurations to the chip. The test's `write_to_hardware()` method usually just calls the same named method on all its test components. This is because the actual register writes will occur in the BFM or driver. One exception is when you are writing a direct test, and it's easier just to write the registers at the test level.

The `testbench::start()` method, if it knows which interfaces and features are in use, starts up all the BFMs, monitors, and drivers. Depending on your architecture, it may also start the generators and checkers.

The `test::start()` method usually just calls the `start()` method on all its owned test components.

The `wait_for_completion()` methods in the test and testbench are used to pause the verification system until the test is finished. Although there are many ways to do this, the examples in this handbook just allow the checkers to say when the test is completed.

The `report()` method in both top-level objects reports their status. For the testbench, it is usually appropriate to report the configurations selected. For the test, it usually just calls the test components.

That's it. This may seem like a lot of methods to write, but you probably do not need to perform tasks in all the methods. Later in this chapter, we will talk about the order in which you might want to implement these methods.

The Test Component Dance

Did you notice that most of the time the test just called the same named methods on the test component? That's because verification has a fractal structure, with repeated patterns. The top-level dance is repeated, with a few changes, in the test. This time, instead of `verification_top()` calling the steps, the test does. The `test_component` also plays a role, subdividing the `start()` method into several lower-level methods, as shown in the following figure.

The `run_component_traffic_()` method has a standard implementation, which calls `randomize()` and then `generate_()`. The `randomize()` method has the same purpose it had for the top-level components: to randomize your random variables. The next method called,

Test Component Dance – Detailed Flow

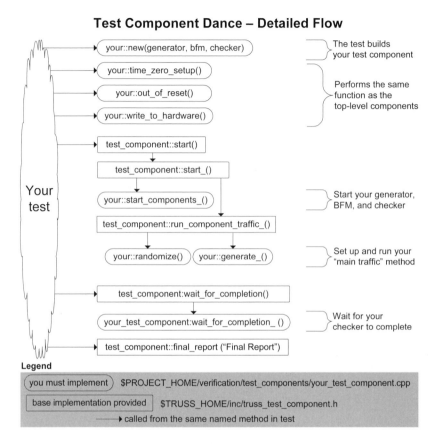

your::new(generator, bfm, checker)	The test builds your test component
your::time_zero_setup()	
your::out_of_reset()	Performs the same function as the top-level components
your::write_to_hardware()	
test_component::start()	
test_component::start_()	
your::start_components_()	Start your generator, BFM, and checker
test_component::run_component_traffic_()	
your::randomize() your::generate_()	Set up and run your "main traffic" method
test_component:wait_for_completion()	
your_test_component:wait_for_completion_ ()	Wait for your checker to complete
test_component::final_report ("Final Report")	

Your test

Legend

(you must implement) $PROJECT_HOME/verification/test_components/your_test_component.cpp

[base implementation provided] $TRUSS_HOME/inc/truss_test_component.h

⟶ called from the same named method in test

generate_() picks up the results of the randomization and interacts with the generator to exercise a feature or an interface of the chip.

Now, it may seem strange that these methods are implemented like this. However, the idea is to separate the various concerns of the test component: starting, randomization, and generation. This, as will be discussed in Part III of the book, creates more adaptable and less brittle code. The organization also sets up the *irritator*, making the transition from a fixed test to an irritator relatively painless.

The Irritator Dance

The `irritator` is an inherited class of `test_component`. Its purpose is to generate background "noise" while the test concentrates on some specific area of the chip. In some sense, using irritators is a way to emulate the real world, where many of a chip's features and interfaces are used simultaneously.

So what does an `irritator` add to or change from the `test_component`? Only one method is changed, and two methods are added. All these changes involve the new `run_traffic()` method, shown in the figure below.

The Irritator Dance – Detailed Flow

| from your_test::start() |
| test_component::start() |
| test_component::start_() |
| your::start_components_() | → Start your generator, BFM, and checker |
| irritator::run_component_traffic_() |
| until irritator::stop_generation() | → Called by test |
| test_component::run_component_traffic_() |
| your::randomize() your::generate_() | → Set up and run your "main traffic" method |
| your::inter_generate_gap_() | → Pause the generate loop |

Legend

| you must implement | $PROJECT_HOME/verification/test_components/your_irritator.cpp |
| you may implement | $TRUSS_HOME/inc/test_component.h, and irritator.h |

The `irritator` overrides the `run_component_traffic_()` method from the `test_component` base, and calls the base class `run_component_traffic_()` method in a loop. This is the nature of an `irritator`: it just keeps on going until told to stop. The method that stops the loop is `stop_generation()`, which is usually called by your test once the main feature or interface has finished being tested. This will be shown in detail in the Part IV of this handbook.

One method you will have to implement is `inter_generate_gap_()`. This method may be empty, for a couple of reasons.

- Your channel has a limited depth, and this limit is used to apply back-pressure to your system.
- Your generator has a built-in delay of some form.

In this handbook we use the checker to throttle the system—because we want to keep a certain amount of data in flight, and the checker is the only agent that knows what has been generated and what has been received. (The chip can handle an unlimited number of back-to-back transactions.)

That's all there is to building an irritator. Note that you will probably start with a test component, and then evolve it into an irritator. It will probably be many weeks into your project before the first irritator is built, but for coverage and finding congestion bugs, irritators are a good choice.

In fact, your first test will probably be even more rudimentary. This first test is the focus of the next section.

Compiling and Running Tests

The sections above described the main building blocks of Truss. The following chapters, as well as later examples, will show how these still somewhat abstract concepts get implemented for real projects. However, before we start looking at more concrete examples, there is one more problem to consider: that of compiling and running a verification environment.

All verification environments need some type of run script to compile and build both the RTL and verification code. In a large project this is not a simple task, because one must track a lot of code, as well as many tools and options.

A goal for Truss is to provide a production-grade run script and makefiles as open-source components. At the moment, a reasonable run script and rudimentary makefiles are provided. They are a good starting point for a run script and provide enough functionality to handle the examples in this handbook. It is the authors' hope that through community effort, these scripts can be fleshed out into something better.

Truss run script

The Truss run script controls which files are compiled and run. It is written in Perl and has a number of switches that controls its actions. The script will first compile all the C++ files, then compile all the HDL files, then link all files into a single executable, and finally launch the simulation. After the simulation finishes, it checks the status of the test run. (This script is used to build and run all the examples on the companion CD.) The script is written in Perl, and can be located at $TRUSS_HOME/bin/truss).

Truss uses some environment variables to "understand" its environment. By using environment variables (instead of .tool_rc files, for example), the system's assumptions are both obvious and flexible. Truss uses only a small number of environment variables, as listed below.

Variable	Function
SIM	Simulator name (such as ncsim, mti, aldec, or vcs)
SIMULATOR_HOME	Path to the simulator install area
TEAL_HOME	Path to Teal's source files
TRUSS_HOME	Path to Truss install area
PROJECT_HOME	Path to top of the current verification project

The file named setup in each of the bin subdirectories of each example on the CD has default values for the TEAL_HOME, TRUSS_HOME, and PROJECT_HOME environment variables. You'll need to set SIM and SIMULATOR_HOME as appropriate for your environment.

Switches

The Truss run script has a number of switches to control its execution. Below is a table that expands on descriptions of the most important switches.

Switch	Function
--help	Prints longer help message
--test <test_name>	Runs the $PROJECT_HOME/testcases/<test_name> test.
--clean [options]	Cleans appropriate selection of the system. Default selection is USER. The following options are available: LOGS - Deletes simulation log files CPP - Deletes user-compiled C++ code HDL - Deletes user-compiled HDL code USER - Deletes all user-generated code (LOGS, CPP, HDL) TRUSS - Deletes compiled Truss files TEAL - Deletes compiled Teal files ALL - Deletes all of the above This switch can be repeated (--clean CPP --clean HDL)
--simulator <SIM>	Selects appropriate simulator from supported list. If switch is not used, then run script reads $SIM. If neither $SIM or --simulator is used script will fail.
--seed <seed value>	Sets random seed to integer <seed value>
--run <number>	Runs the selected test a number of times
--sim <sim>	Builds and runs using <sim> as the simulator.

For a full description of all switches from a command line, run the following:

```
$TRUSS_HOME/bin/truss --help
```

The Truss makefile

As is customary in the coding world, a makefile is used to build the objects and archives.[1] The Truss makefile may be a good starting point for your makefiles. Almost all of the directories in the examples include a standard makefile, located in the `/inc/Makefile` subdirectory of Truss. As with the `truss` script, this makefile is both too simple and too complex.

The makefile for three sources is shown below.

```
STATIC_LIB = $(LIB)/directory_name.$(SIM).a
INC = -I../a_referenced_directory
SRCS = \
    $(SRC)/file_one.cpp \
    $(SRC)/file_two.cpp \
    $(SRC)/file_three.cpp

include $(TRUSS_HOME)/inc/Makefile
```

The first line identifies the output static library name. The next three lines identify the source files. The last line includes the standard makefile. Most makefiles follow this form.

The Truss makefile has all the compiler switches to build the sources for a variety of simulators.

The First Test: A Directed Test

Because starting something new is not always easy, this section helps make the process easier by addressing how a first test can be written in Truss. This section concentrates on the steps you need to do, and how a test can be built up from scratch. The next chapter shows a complete first example and focuses more on the flow.

Your first test will probably be a simple directed test, with a `test_component` that does not have a generator and possibly not even a checker. It will probably interact directly with the BFM or driver.

[1.] The final shared object is built by the `truss` script.

Focus your initial efforts on the driver and BFM. Write a "first cut" at the driver class, making it have the methods that seem right to you. You may or may not need a monitor, depending on the protocol or feature to be tested.

Next, create a testbench that includes that driver/BFM and think about how to get clocks to the chip and get it out of reset.

Now make a test class and get the whole thing compiling. Before moving on to connecting the test to the driver with a test component, make sure the chip is cleanly out of reset, as this can be done by the testbench's `out_of_reset()` method.

The next step is to make a simple `test_component`. This component will probably just be a directed exercise, with perhaps a few reads and writes or just a few calls to the driver. Note that you may use the `test_component` pre-implemented methods if you are comfortable with them, but for a first test it might be better just to override the `start()` method directly. This is because that's easier than remembering where to put your randomization and traffic-generation code.

If there is any configuration, use the chip's default configuration. Don't try to randomize anything yet.

Doing the checking can be tricky, so let's worry about that last. We'll probably be looking at waveforms for the first few days anyway.

Now build a test that has your `test_component` as a data member. Initially, have the test call the same named methods on your test component.

Note that the `wait_for_completion()` method probably just returns, if you implemented the `start()` method. However, if you used the `generate_()` method of the standard `test_component`, you'll want to trigger a `condition` variable at the end of your `generate()`. Then, the `wait_for_completion()` would just wait for the signal to be triggered, as shown below:

```
class your_test_component {
  //...your other code here...
  private:
  teal::condition done_;
}
```

Then, in the last line of the `your_test_component::generate_()` method, do this:

```
void your_test_component::generate_()
{
  //...your directed exercise code here...
  done_.signal();
}
```

Then your `wait_for_completion_()` would look like this:

```
void your_test_component::wait_for_completion_()
{
  done_.wait ();
}
```

That's it! You have created your first Truss-based test.

The Second Test:
Adding Channels and Random Parameters

Software engineers count "one," "two,"—and then "many." This is because only the first three times they use a technique are significant. After that, everything looks like "many." By writing the first test, we've counted "one." Now we will count "two." The next section will cover the "many."

In this, the second test, we'll get more sophisticated. We'll add the agent layers and also add the generator and checker. These are the steps you need in order to create more advanced, randomized tests. You will probably create several directed tests before you need these additional features, but because this is a book we need to keep moving along.

Remember that the generator and monitor generally have pure virtual methods to communicate the results of their work. We'll add our agents to these methods. There will be an agent for the generator, the driver/ BFM, the monitor, and the checker. Why all this complexity? Because there are many interconnection techniques, each one involving some architectural trade-offs. These trade-offs are talked about at length in the OOP Connections chapter in Part III of this handbook.

To make the connection between the agents, we'll use a Truss channel. So let's digress a bit and look at a channel.

The channel classes

Verification systems have a lot of producer/consumer relationships. For example, a generator can be considered a producer and a BFM considered a consumer. However, it is a good idea to minimize the knowledge and assumptions of the interface between these two loosely cooperating objects. One way to decrease the coupling between these components is to use an *intermediary object*. An intermediary object would allow the two communicating objects to be anonymous or separated in time. The concept behind this object is called a pipe, mailbox, or channel. Truss uses the term *channel*.

Truss separates the roles of producer and consumer by having two abstract base classes, `channel_get` and `channel_put`. This clarifies the roles of the two communicating objects. For example, the constructor of a generator would take in a `channel_put` class, because it puts data into a channel. A monitor's constructor would also take in a `channel_put` class. A BFM or checker's constructor, on the other hand, would take in a `channel_get` class.

In Truss, the `channel_get` and `channel_put` classes are templated. This is one of the very few places where Truss uses templating. Don't worry if you don't understand all the details of the code in a first read-through; these templates are not hard to use, as will be shown in the chapter discussing a single UART example.

The reason we use a template is to encourage a strongly typed channel. Another reason is to allow you to have the choice of using pointers or actual objects and data in the channel. In general, the authors use pointers to objects only when virtual methods are needed. Otherwise, we put the objects themselves in the channel. This simplifies any memory-management issues.

The `channel` class joins the two concepts `channel_put` and `channel_get`. This class adds the storage for the actual data, as well as the signaling and mutual-exclusion mechanisms. In addition, `channel` also supports a `depth` concept, for designs that want to implement back-pressure in that way. The interface for a `channel` class, as well as the base classes, are in `/truss/inc/truss_channel.h` on the CD.

The `channel` class also provides for other `channel_put` objects to be attached to a channel. This allows the data of one `put()` to be replicated across many channels. The common use for this is when a generator creates a data item and both the checker and BFM should get the data. It is also useful if there are multiple listeners to a channel, such as in an Ethernet broadcast, or multiple monitors for a data interface.

Building the second test

Now that we have channels, let's use them for the agents. This section is a bit high level, because every situation is different. We'll give general direction, but after you read this chapter, take a look at the next chapter for a first complete example.

Let's say that you are working on a chip interface called `my_interface`. You might have a generator that looks like this:

```
namespace my_interface {
  class my_data;
  class generator {
    public:
      void generate(); //make one
      virtual void post_generate_(const my_data&) = 0;
  }
}
```

We are concerned with the `post_generate_()` method. This is a pure virtual method, so we must implement it in our inherited class. Let's assume we want to add a channel interface, like so:

```
#include "generator.h"
namespace my_interface {
  typedef channel_put <my_data> generated_channel;
  class generator_agent: public generator {
    public:
      generator_agent(generated_channel* out)
                                       : out(out) {}
```

```
      virtual void post_generate_(const my_data& d) {
        out_ ->put(d);
      }
  private:
    generated_channel* out_
  };
};
```

By building a `generator_agent`, we have abstracted how the generator gets the created data to the driver/BFM.

A similar situation exists in the monitor:

```
namespace my_interface {
  class results;
  class monitor: public truss::thread {
    public:
      void start();
      //the connection method
      virtual void data_received_(const results&) = 0;
  };
};
```

And likewise for an agent for the monitor:

```
#include "monitor.h"
namespace my_interface {
  typedef channel_put <results> generated_channel;
  class monitor_agent : public monitor {
    public:
      monitor_agent(out_channel* out) : out_(out) {}
      virtual data_received_(const results&) {
        out_ ->put(r);
      }
    private:
      out_channel* out_;
  };
};
```

But what about the other side of the channels? These objects are the `driver_agent` and `checker_agent`, respectively. Their job is to take the data out of a channel and act on the data.

Remember, we are discussing channels here because that's how we wanted to implement the agent layer. This could have easily been a more generic producer/consumer model, or an event-driven method, but implement what feels correct for you. (All the examples in this handbook use channels.)

Here are the classes for the driver and checker and the inherited classes for their agents:

```
namespace my_interface {
  class driver {
    public:
      void send_data(const my_data&);
  }

  typedef channel_get <my_data> input;

  class driver_agent : public driver,
                          public truss::thread {
    public:
      driver_agent(input* drain) : drain_(drain) {}
      //must have a start to drain the channel
      void start_() {
        for (;;) {
          send_data(drain_->get());
        }
      }
    private:
      input* drain_;
  };

  class checker {
    public:
      check_data(const my_data&, const results&);
  };

  typedef channel_get <results> checker_in;
  class checker_agent : public checker,
                          public truss::thread {
    public:
      checker_agent(generated_channel* generated,
              checker_in* actual) :
              generated_(generated), actual_(actual) {}
```

```
    void start_() {
      //Check the data!
      for (;;)
        check_data(generated_->get(),
                   checker_in->get());
    }
  private:
    generated_channel* generated_;
    checker_in* checker_in_;
  };
};
```

The authors realize that there is a lot of code to look at, but just skim it over to get the general idea. The general technique is to inherit a class, add a channel, and append _agent to the name.

After the agents have been built, they should be added to the testbench. The testbench holds the generators, drivers, monitors, and so on. The test, on the other hand, holds the test components.

Building the second test's test_component

The test_component is relatively straightforward. A test_component constructor takes in the parts of the testbench you need. Remember, the entire testbench is not taken as a parameter, because then we would have to make assumptions about the name of the parts we needed. Also, by taking in only the parts we need, several of our test components can be used in the same chip.

The most likely candidates for the constructor's parameters are the generator, the driver, and the checker.

The rest of the test component usually just forwards its calls to the appropriate objects. An example test component is shown below.

```
namespace an_interface {
  typedef class bfm;
  typedef class generator;
  typedef class checker;
};

#include "truss.h"
```

```
namespace an_interface {
class a_test_component : public truss::test_component {
  public:
    a_test_component(const std::string n,generator* g,
                     bfm* b, checker* c);

    virtual void time_zero_setup() {
     bfm_>time_zero_setup();};
    };
    virtual void out_of_reset(reset r) {
      bfm_->out_of_reset(r)};
    };
    virtual void randomize() {/* next section */;}
    virtual void write_to_hardware() {
      bfm_->write_to_hardware();
    };
  protected:
    virtual void generate() {generator_->generator();};
    virtual void wait_for_completion_() {
      checker_->wait_for_completion();
    };
    virtual void start_components_() {
      bfm_->start(); checker_->start();
    };
  private:
    generator*  generator_;
    bfm*        bfm_;
    checker*    checker_;
  };
};
```

Although your actual test component will be a bit different from the code above, the general form will probably be the same.

Adjusting the second test's parameters

As soon as you introduce randomization into a test, you'll probably want some knobs to control the randomization. Sweeping most parameters through an entire integer range would chew up a whole lot of simulation time. Besides, it's probably either (1) not interesting, or (2) unacceptable to the register associated with the integer.

A knob is a technique that uses other variables to control the range of a random variable, either directly or indirectly. In this example we'll concentrate on controlling the random variables directly. (The examples in the handbook use the Teal dictionary feature to pass parameters from a number of sources to the method that will use the knob variables.)

For example, consider a test for a CPU. Assume that a `cpu_generator` class has a `send_one_operation()` method that is called by a `test_component` to tell the `cpu_generator` to create one random operation. The generator is guided by dictionary variables. It is best to put the variables to randomize in a separate function at the top of the source file, because the seeding depends on line number. That way, the sequence of values selected does not change if the code below is reorganized. Of course, new random values chosen will be different for each master seed.

Here is an example function for generating the `operand_a` variable of a CPU operation:

```
namespace {
  uint32 get_operand_a(uint32 min_v, uint32 max_v) {
    uint32 returned; RAND_RANGE(returned, min_v,
      max_v); return returned;
  }
};
```

In the `cpu_generator`, the following lines could be used:

```
static uint32 min_operand_a =
  dictionary::find(name_ +    "_min_operand_a", 0);
static uint32 max_operand_a =
  dictionary::find(name_    + "_max_operand_a", ~0);
operand_a = get_operand_a(min_operand_a, max_operand_a);
```

This same style is used for the other operand and the operator variables.

So who sets the knobs? There are four ways: (1) use the default specified in the `dictionary::find()` call as the second parameter; (2) put the knob value on the command line; (3) use a knob configuration file; or (4) (finally) write code to use the `dictionary::put()` call, which is the mechanism used in our example. Note that because the Teal dictionary is used, both the command line and the knob file can be added later without the need to modify any of the example code.

The test constrains the test component with respect to the number of times the generator is called. Of course, this specifies the number of operations sent to the arithmetic logic unit (ALU). The code is shown below.

```
teal::dictionary::put(test_component_->name +
  "_min_operations", "4",
    teal::dictionary::default_only);
  teal::dictionary::put(test_component_->name +
  "_max_operations", "10",
  teal::dictionary::default_only);
```

Note that the name of the `test_component` is used. This allows the test to pick any name for the `test_component` and still have the code work.

However, be careful with the spelling of the knob variables. They must be spelled the same in both the `find` and the `put` routines in order to make a connection.

Now that the randomization and knobs are connected, we have completed writing the second test. In some ways, this test is rather sophisticated. It uses the Truss framework, and adds agents by using channels to connect the wire-layer classes to the transaction-layer classes.

The testbench created and wired up the generator, driver, monitor, and checker. The testbench can bring the chip out of reset and start the monitor.

The test itself is rather reasonable. It creates and connects the test component to the generator, driver, and checker in the testbench.

The Remaining Tests:
Mix-and-Match Test Components

So now what do you do after creating this second, more-sophisticated test? You do what we verification engineers always do—create more tests! As these tests are being written, new test components will also be created, some of which could be used in several tests. Deciding which test components to adapt to different tests is the major activity (besides writing more tests) after you have written the first two tests. This is the "many" count that we talked about earlier.

Of course, you'll be doing other test-related activities, such as adding randomness to the existing tests and looking over your verification test plan to make sure you know when you're done.

And how do you go about adapting a test component from one test into another? You could just put the new test component in the test and wait until both of them are completed. However, as explained in the Truss Basics chapter, there is another way: use the Truss concept of irritators, and warm over, or "recrystallize," the existing test component to an irritator.

Converting the test components to irritators usually just involves deriving the existing test component with the `truss::irritator` component. Then, the appropriate methods will be overridden and the only method you have to write is `inter_generate_gap_()`. There are many ways to implement a gap, from the simplest (pausing a number of clock cycles), to the more complex (using back-pressure and bursty traffic). If the checker were inherited from Truss's checker, you can also just wait for generated data to be checked.

This process of writing a new test continues for all the rest of the features and interfaces of the chip. Remember, the more irritators a test has, the more likely it is to model what actually happens when the chip design is realized in silicon.

Summary

This chapter tried to clear the fog of how to go about using Truss. We started with a review of the top-level dance, and then showed that the dance also existed in other layers of the system.

We looked at the tools provided by Truss, which are the `truss` execution script and a standard makefile.

We covered writing the first test, concluding that it will probably be a directed test. Then, we took the test up a notch, adding connection agents to the generator, driver, monitor, and checker. We introduced the Truss channel as the interconnect technique, but noted that there are many other techniques.

We looked a bit at control knobs, a technique for passing parameters to constrain randomization. (There are many techniques for constraining random-variable generation.) This chapter showed how to harness Teal's dictionary to hook up bounds for randomization.

We finally discussed what to do after the second test. The idea is to write more tests for that interface or feature, and also test the rest of the chip. The key part of writing more tests is to keep an eye out for what you can "steal" (rather, "adapt") for other tests. By creating irritators, you can use the functionality of other tests as background activities. In this way, the chip is stressed more—and more faults are found prior to production.

Truss Example

I know that you believe you understand what you think I said, but I'm not sure you realize that what you heard is not what I meant.

Robert McCloskey

Coding is tricky, because we take the great ideas, techniques, and trade-offs and actually make decisions. We put fingers to the keyboard, and decisions are made and trade-offs are fixed in code. Furthermore, learning a new technique only makes the coding task more difficult. An example, or several examples, can help put the technique into perspective.

This chapter is the first example of how to use Teal and Truss in a verification system. It's useful to build and run some example code when learning something new. So, install the code on the CD and noodle around with it a bit. You can add printf's and change the code a bit.

If you want, use this chapter as a guide to some of the more interesting parts. This chapter is not quite a map to the "homes of the movie stars." Instead, it is more like a mariner's map. It helps you to navigate in tricky waters.

Overview

This chapter provides a first complete example of using Truss, where you can actually compile and run the code. The code is not as complex as what you would encounter in a fully featured chip. However, all the main parts are here to consider. The source files may seem silly or overly complex for the chip we are trying to test, but we are trying to demonstrate how to structure a verification system for a real project. Your chips will have plenty of complexity to manage.

This chapter does not walk through every code file. We are all capable of reading code. What it does instead is look at some of the more important aspects of the verification system.

Directory Structure

In order to help you navigate the source files, it's good to show the main directories that comprise a Truss-based system (shown below). We've also included only the main files we will be working with.

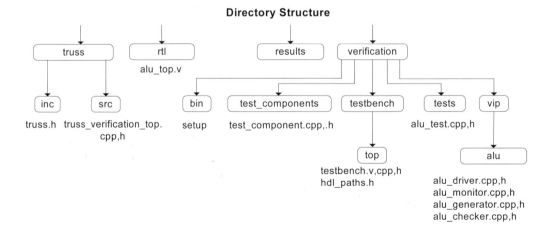

Directory Structure

The source code for the chip is in the /rtl directory. How does the truss run script know this? The file /verification/testbench/top/ hdl_paths.vc is used to specify the paths to the RTL and the RTL include directories. This is so that the RTL files can be rooted in a place different from the verification directory.

The /results directory is where you run the tests from. It also can be wherever you want. The authors generally put this directory in some non-backed-up networked storage area that is independent of the source-code control system. In the handbook example, the /results directory is placed in /examples/alu, at the same level as the /verification directory.

The /verification directory contains all the source code for the verification system. The /bin directory is there for the project's local scripts. The authors usually put a setup script there and alias setup to it.

The other four subdirectories—
/tests, /testbench, /vip, and /test_components
—are where the actual source files are. The /tests directory is where your test_name.cpp and test_name.h exist. These files are used when you give the --test <test_name> option to the truss script. (Use the --config option to the truss script to select a directory.)

The /testbench/top directory contains the C++ and HDL sources for the top-level testbench. If you have more than one chip in your simulation, it may be useful to have /testbench/<chip_name> directories.

The /vip directory is where chip interface classes go. There should be a subdirectory for each interface and major feature you need to test. The idea is that the code in these directories is fairly portable, and may contain purchased VIP as well as project and company-created VIP. In our example, there is only the /alu directory.

The /test_components directory contains the scenarios that you want to run. For this example, we'll only run one scenario, called test_component.

Theory of Operation

· ·

We'll be testing a really basic ALU chip. It takes in two 32-bit operands and performs a simple logic or arithmetic function. We'll use a legacy c-model for comparison with what the chip produces. The output consists of a result and an "operation complete" status interrupt. The `test-bench.v` will instantiate this ALU module and provide system clocks for the chip and verification system.

The main objects are shown below.

ALU Example: Objects and Connections

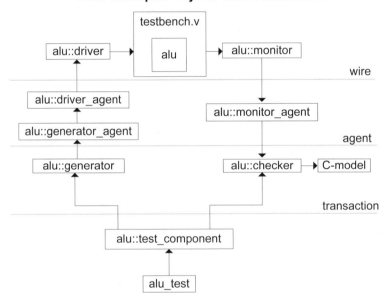

Because there is only one interface in the chip, we'll just refer to the components by their functionality. In other words, we'll say "driver," although in a chip with many drivers we would need to say which interface we are talking about. (Note that we do scope the code in an ALU namespace.)

In the testbench C++ class, we have all the components of the ALU interface layer. There is a wire-layer driver and monitor, with their accompanying agents. There is a generator and a checker. The checker is interesting, because we have a legacy c-model of the chip, which will be used by the checker.

There is also a `test_component` class, which runs a random number of operations through the chip. And, of, course, there is a `alu_test` class, which builds a `test_component`, giving it the generator, checker, and driver from the testbench.

The following illustrates the wires used by the verification system:

ALU Example: Wires and Objects

The driver and the monitor take care of the protocol into and out of the chip. The testbench takes care of bringing the chip out of reset.

The remaining sections highlight some specific "points of interest" in the code. The code itself, being the first example, is not that big. If you want to follow the code through its execution, start with the Truss `verification_top.cpp`, then move on to `testbench.cpp` and `test.cpp`.

Running the Simple ALU Example

You might want to see the log messages on the screen, so let's talk about how to run the example. In the `/examples/alu_tutorial/bin` directory, there is a setup script. If you look at the `setup` file, it sets up a few environment variables that are needed by the run and make tools.

First, source the `setup` file, then execute the following:

```
$TRUSS_HOME/bin/truss --test tutorial_test
```

The `truss` command has many more options; type `truss --help` for a synopsis.

You should see the C++ source files being compiled, and then the test should run. When the test runs, a series of printouts will announce the flow through the test. Remember that, by default, Teal prints the file and line number of the message.

Points of Interest

The next few sections address specific places in the code. These sections follow the general way you go about hooking up a chip to a Truss-based verification system.

For example, the first thing to be concerned with is bringing the chip out of reset. After that, you'll probably want to pick an interface and write the driver and monitor classes. Then, you decide upon some specific operations you want to perform and write the test component to exercise the interface or feature.

In general, the test builds the test components and ends when the last operation completes—that is, when the test component's `wait_for_completion()` returns.

Hardware Verification with C++

Power-on Reset

Most chips have a power-on reset sequence. This sequence can be basic, or rather complicated. In this example we address a basic sequence.

The chip has a *reset* line, which is pulled low to initiate a reset. After the line is asserted, the chip performs its reset sequence. This chip only needs a fixed-duration pulse.

The `testbench` class is responsible for bringing the chip out of reset. The testbench methods `time_zero_setup()` and `out_of_reset()` are called by the top program to bring the chip on line. In our ALU example, we'll use a reference clock to count a number of cycles to keep the `reset_n` low.

Below are the snippets of code that perform the chip reset. The methods are located in `testbench.cpp`. Note that the `verification_top()` provided a path to the top of the testbench; the path is stored in the variable `top_`.

This method is called first by `verification_top()`:

```
void testbench::time_zero_setup() {
  teal::vreg reset(top_ + ".reset");
  reset = 0;
}
```

Then, this method is called:

```
const teal::unit32 reset_count = 10;
void testbench::out_of_reset(reset r) {
  teal::vreg reset(top_ + ".reset");
  teal::vreg clock (top_ + ".clock");
  reset = 1;
  for (int i(0); i < reset_count; ++i) {
    teal::at(teal::posedge(clock));
  }
  reset = 0;
}
```

That's all there is to it. Now the chip is ready for operation.

Driver and Monitor Protocol

Now that the chip is out of reset, we can start to drive it. This chip has a simple protocol for sending operations to perform. Assuming op_done is asserted, the driver puts op_code, operand_a, and operand_b on the wire. Then it asserts do_op and waits for op_done to be asserted. The code to do this is in alu_driver.cpp and is shown below:

```
void alu::driver::send_operation(const operation& op){
  op_code_ = op.code;
  operand_a_ = op.operand_a;
  operand_b_ = op.operand_b;
  op_valid_ = 1;
  at (posedge(op_done_)); //wait until accepted
  op_valid_ = 0;
  at (negedge(op_done_));
}
```

The variables above with the trailing "_" are data members and are Teal vreg objects that are connected to the chip.

The monitor code is fairly simple as well. The monitor uses a Truss utility thread class called run_loop. It consists of two methods, loop_condition() and loop_body(), which are run in a separate thread. The idea is that a large number of monitors are infinite loops of "wait for trigger" and then "gather data." This class represents that concept.

The loop_condition() method of the monitor waits for op_done to go high. The loop_body() method then copies the result into a local variable. It then calls the pure virtual method operation_completed() to connect to the monitor agent. Here is the code, in cpu_monitor.cpp:

```
void alu::monitor::loop_condition_()
{
  at (posedge(operation_done_));
}

bool alu::monitor::loop_body_()
{
  receive_completed_(result_.to_int());
  return true; //continue loop
}
```

Other than the reset logic (and the watchdog timer), the monitor and driver are the only code to interact with the chip wires.

Next we'll look at how we come up with the operations to be sent to the driver.

The alu_test_component

We now run a random sequence of operations through the ALU, testing the basic operations with random operands. The test_component::start_components_() method is used to run this exercise.

The code is shown below.

```
void alu::test_component::start_components_()
{
  driver_->start();
  checker_->start();
}
```

Like most test components, this one just starts the lower-level components.

Checking the Chip

Because we do verification for a living, the automated checking of the chip's results is important. In our case, we have a legacy c-model of the ALU and will use it to check that the answer is what we expected. The checker waits for the monitor agent to deliver a completed operation. Then it uses the inputs sent by the generator to have the c-model come up with the expected result.

The c-model prototype is shown below.

```
#if defined(__cplusplus)
  extern "C" {
  #endif


  unsigned int alu_model(unsigned int a,unsigned int b,
                         unsigned char op);


#if defined(__cplusplus)
  }
#endif
```

Note that the `ifdefs` allow the code to be compiled by both C and C++ code.

This key algorithm is in `checker.cpp` and is shown below.

```
void alu::checker::start_()
{
  for (;;) {
    operation gen = generated_->get();
    teal::uint32 actual = actual_->get();

    if (alu_model (gen.operand_a, gen.operand_b,
                   gen.op_code) == actual) {
      log_ << teal_info << " EXPECTED: sent " << gen
           << " == " << actual << endm;
    }
    else {
      log_ << teal_error << " sent " << gen
           << " != " << actual << endm;
    }
    if (!generated_->size()) {
      completed_flag_.signal();
      return;
    }
  }
}
```

The checker works fine as long as the `operation_done` is in synch with the result. However, the checker can be wrong if the monitor misses a result or somehow inserts an extra one. We could have registered the chip inputs at the same time as we got the results. However, by doing this we make the assumption that there are no queuing or pipe stages in the ALU. This assumption works fine for our example, but it is probably not valid for most ALUs.

Completing the Test

. . . .

When does the test stop? When `verification_top()` calls the test's `wait_for_completion()`, which in turn calls the test component's `wait_for_completion()`.

In turn, the test component's `wait_for_completion()` calls the checker's `wait_for_completion()`. The authors agree that this sounds silly, but in the later examples we actually do a bit more than just forward the call.

In the end of the forwarding chain, it's the checker that actually decides when the test is done. This makes sense, because the checker is the best able to "judge" what the chip did and when all the inputs have been checked.

But how does the checker know? There are many possible ways, but in this example the checker assumes that when the generated data channel runs dry, the test is over. This is a valid assumption—as long as you make sure that the generator can always be one step ahead of the checker. (If your chip has any latency, this is not a hard assumption to sustain.[1])

The checker code is shown below—

```
void tutorial::checker::wait_for_completion()
{
  completed_flag_.wait();
  //note that the checking thread completed normally
  completed_ = true;
}
```

—and at the bottom of the main check loop:

[1.] Note that an intergenerate delay should not affect when the expected data are sent to the checker. The point is that even when delays are inserted, this model should be valid.

```
if (!generated_->size()) {
  completed_flag_.signal();
  return;
}
```

Remember that after the `wait_for_completion()` returns, the top calls the `report()` method in the test. The test calls the test_component's `report()` method, which in turn calls the checker's `report()` method.

The `report()` method prints the state of the `completed_` boolean. In this way, when you have multiple test components and the watchdog timer shuts the simulation down, you can tell which checkers have not completed.

Summary

This chapter is a tutorial on the Truss framework. We exercised a simple ALU, but implemented all the parts of a Truss-based verification system. The main objects and their connections were shown. The directory structure was introduced so we can find our way around the code. Then, the chip and the HDL connections were shown.

After laying out the verification system and showing how to run the example, we looked at how the chip was to be brought out of reset. We did a quick side tour to talk about how to run the example. Running the example produces many log messages, but this is probably a good thing when one is learning.

We showed how to bring the chip out of reset and how the driver and monitor interfaced with the chip. One point to note is that while this interface consisted of only a few wires, many interfaces in real protocols are this small. Of course, your code will be more detailed.

We looked at an important part of the verification system, the checker. In this example, the checker used a c-model to check that the chip was working correctly.

The last thing we looked at was how the test stopped. We looked at the normal path, ignoring the watchdog timer. We showed how the checker was in charge, pausing the end of the test until all the generated data had been checked. The interesting point to note is that the checker may have had errors, but it will continue until all generated data have been checked. The Truss utility class `error_threshold` can be used to terminate the simulation in the case of excessive errors. The Truss `verification_top()` does this.

Whew! We made it through the first example. Time for a coffee break and some foosball!

Part III: Using OOP for Verification (Best Practices)

This part of the handbook explores what it means to write OOP-based code. It's not easy to "get it" when it comes to OOP. There are many techniques, and experience plays an important part.

We'll walk through the activities of programming, showing examples and experiences that form the design and coding biases often found in OOP-based verification systems.

We'll end each section with a short sentence about the lesson learned from each example or experience. This is in no way meant to be a rule. Rather, it's another trick, to be added to your bag of tricks you can use— or not—as appropriate.

This part addresses the following themes:

- The shift in thinking that usually occurs when you start working with OOP

- How to bias a design towards managing complexity when coding with OOP

- Techniques useful in making classes and connecting them
- Code techniques useful in writing OOP-based code

Thinking OOP

NOBODY expects the Spanish Inquisition!
Amongst our weaponry are such diverse
elements as fear, surprise, ruthless
efficiency, an almost fanatical devotion to
the Pope, and nice red uniforms—Oh damn!

Monty Python, episode 15, 1970

Getting your brain around OOP is a challenge. You may have followed the syntax of classes, inheritance, and so on. But when should you write classes or use inheritance? What about templating and operator overloading? A little befuddlement is okay—OOP requires a shift in thinking, and mental fog is a natural result.

This chapter will get you "thinking OOP." The reason OOP is all muddy is that there are no rules. "Thinking OOP" is more about using a set of coding biases and lessons learned than in making trade-offs. Sure, we could have pretended there were rules, providing numbered steps such as, "first you must blah, blah, blah," or "you must always apply by blah, blah, blah," but no one would remember. Instead, this handbook tries to teach you how to ride the "OOP bicycle." Learning to "think OOP" is not trivial, but once you've learned, you never forget.

Overview

We now introduce thinking and using OOP in stages. From the first stage (the "big picture") to the last (coding), we introduce an "arsenal of weaponry" that has proved useful for programmers. This arsenal requires a few chapters. In this chapter we concentrate on framing the OOP process. We talk about the difficulties in managing complexity and creating adaptable code. We then discuss the difference between the interface and the implementation of a piece of code. Subsequent chapters cover architecture and coding.

Remember, verification is neither simple nor easy. Any serious attempt to verify hardware will result in a complex system. Consequently, it is important to realize that the complexity of a verification system is not the result of poor implementation, but is largely intrinsic to the problem of verifying a complex design.

Object-oriented programming in C++ was developed to help manage complex problems,[1] not eliminate them. The goal is to make the complex appear simple without introducing unexpected behavior. The trick is to keep the focus on making things seem as simple and clear as possible, while minimizing the use of "magic" code or confusing connections. This will nonetheless create a bit of a conundrum, as what is simple and clean to one is often perceived as unnecessarily complex and "sneaky" by another.

There is no "silver bullet" to slay the werewolf of complexity. Verification complexity needs to be managed differently across different types of projects. For example, System-on-a-Chip (SoC) designs are complex because they often involve several independent input/output (I/O) subsystems. SoC and graphics chip designs are complex because of pipelined and interrelated processing. The best solution to complexity is communication, through understandable design and code (abstractions, minimal assumptions, and so on), combined with a drive toward common-sense simplicity.

---

[1] In an ironic twist of fate, "C with Classes," the progenitor of C++, was invented to solve simulation problems.

Furthermore, making the design and code adaptable adds to the difficulty of verification—yet it is exactly this additional difficulty that OOP was created to manage. In a fast-paced and increasingly complex product cycle, writing adaptable code is as important as managing complexity. The concept of creating and adapting code seems simple enough, although in practice it is very difficult, for a number of reasons. This chapter looks at some of the reasons why building adaptable code is difficult. Don't get discouraged; often adaptable code is a natural by-product of a well-reasoned design.

One way to think about managing complexity and creating adaptable code is to look at what is holding us back. In the real world of verification development, there are rarely perfect solutions. Nevertheless, we can build systems that make appropriate trade-offs. To this end, some sections of this chapter include a table of trade-offs to help you make the most appropriate choice for your code.

Sources of Complexity

When you sit down to write code, there are several constraints that slow down the coding process. These constraints can be viewed as adding complexity, because they make an inherently difficult problem even harder. Some of these complexities arise from external sources such as teamwork (that is, local personalities or working with remote sites). We'll touch on teamwork lightly, following our discussion of complexity.

Other complexities are created when a solution is implemented. This is because any solution, almost by definition, involves trade-offs. The authors call this *implementation complexity*, and discuss it in the next section.

Essential complexity vs. implementation complexity

In any verification task there are algorithms and procedures that are required by the specification. In USB, for example, there is a process called enumeration that has a prescribed algorithm for both the host and the device. This is called *essential complexity*, because it is required. When that protocol is implemented in classes and code, some additional complexity is created. For example, the host and device interrupt code must try different scenarios. The authors call these classes and code *implementation complexity*. The classes and code are needed, but are more an artifact of the solution than a real part of the problem. Why is this distinction important? Because you cannot get rid of the essential complexity, the goal is to make the essential complexity as simple as possible, and keep the implementation complexity as minimal as possible.

Implementation complexity is to some degree always created when you are designing or coding essential complexity. For example, although the PCI Express protocol specifies endpoints and a root complex (the host node, or top of the tree), no data structures are specified to manage these concepts. When these are coded in the verification of a root or endpoint, they are implementation complexity.

Remember, engineering is all about building the appropriate solution to a problem, creating problems as a result of that solution, solving those problems, and so on. The successful engineer transforms the big problems into a series of solutions and little problems that are acceptable for the task at hand.

It is important, as much as possible, to use the terms and connections identified in a protocol, chip, or system specification. This will minimize the implementation complexity and provide a basis for a mental model of operation. Try to minimize the implementation complexity, but understand that it will always be present.

> *Be aware of the essential complexity of the problem, and be even more aware of the complexity created by the solution.*

Flexibility vs. complexity

To make a verification system that is flexible also appear simple is exceedingly difficult. Flexibility and complexity are often trade-offs, and usually flexibility wins. It often helps to keep in mind that the clients (users) of the code you write are intelligent, but time-limited. An overly complex solution will do more to slow them down than a simple, but tedious, interface.

As an example, consider a memory subsystem of a verification testbench. Assume that this memory subsystem is on the main bus of a chip. There are many possible questions to ask when designing the interface. For example, should there be separate back- and front-door accesses? Is randomization needed? Should all writes be checked to confirm that the chip has accepted the data? What happens when the chip reads memory that was not initialized? Is this an error, or should it be ignored (and random data or X's returned)?

The following is an example of a possible class interface. How obvious is it that the design questions above were answered in a flexible, but not complex, way?

```
class memory_bus : public verification_component {
  public:
    memory_bus();
    //The zero-time memory access methods:
    teal::reg back_door_read(uint64 address);
    void back_door_write(uint64 address, reg value);

    teal::reg front_door_read(uint64 address);
    void front_door_write(uint64 address, reg value);

    //Will randomly select front or back door every time
    teal::reg read(uint64 address) {
      bool front_door; RAND_RANGE(front_door, 0,1);
      return(front_door ?
             front_door_read (address) :
             back_door_read (address));
    }
    void write(uint64 address, reg value); {
      bool front_door; RAND_RANGE(front_door, 0,1);
      if(front_door) front_door_write (address, value);
      else back_door_write(address, value);
    }

    virtual teal::reg handle_DUT_unitialized_read
                      (uint64 address) = 0;
  };
};
```

There is no immediate solution to the flexibility-vs.-complexity trade-off. The "current best" answer will evolve as your team changes its members and gains experience. The class above certainly seems complete, if possibly a bit too complex. One thing to note is that there is both front- and back-door access as well as a random method. This seems overly complex, as the random method could be implemented in an inherited class if that is what coders want. In this case, that interface should probably be removed.

Now suppose that team members designed their code to work independently of whether the memory read/write method was front or back door. In this case the random method should be the only approach, and the front- and back-door accesses could either be moved into the private access or left to subclasses to implement. Note that by removing the explicit calls to front- and back-door access, we are making the code both less clear and more flexible. This is either good or bad, depending on whether the team wants to write code that is independent of the front- and back-door access method.

Now take a look at the pure virtual method to handle a read to uninitialized memory, `handle_DUT_unitialized_read()`. By making this method pure, an inherited class must be created. However, even this is confusing. Is this the method for a verification-initiated read or a chip read? Consequently, there should be two methods to cover both cases. Also, while a flexible solution requires two methods and an inherited class, it may be appropriate to make a simplifying assumption.

Suppose that the team considered a read to uninitialized memory by the verification system to be an error. This could simply be written into the implementation of the read method. However, the chip side is not so clear, so maybe just returning X's might be the team's preference, and this pure virtual method could possibly be removed.

Here is a modified memory class resulting from the previous discussion:

```
class memory_bus : public verification_component {
  public:
  memory_bus();
  //The zero time memory access methods:
  teal::reg back_door_read(uint64 address);
  void back_door_write(uint64 address, reg value);

  teal::reg read(uint64 address);
  void write(uint64 address, reg value);
};
```

A class interface can be flexible or simple, depending on the specific needs of the verification effort.

Apparent simplicity vs. hiding inherent complexity

One of the goals of good coding is that there should be no surprises when one tries to understand the code. As a counter-example, an interface may appear simple, but in practice it may have a usage model that affects the simplicity of the interface. This often shows up when you try to inherit from a class or call the methods in a different, but rational, order compared to what the original coder intended.

Example: How hiding complexity can create confusion

Here is an example of where "hiding the complexity" actually made the system harder to understand. In verification there are classes that manage the top level of a subsystem, other classes that manage the transmission of data (often called stimulus generators), and still other classes that monitor the output of the chip. One verification team decided to put these concepts into separate base classes that all used a common root class. The common root class had the usual `init()`, `start()`, `,stop()`, and `post_run()` and methods. All of the subsystem classes inherited from these base classes. The constructor of the base class put all the instances on a master list, with an `enum`, called `type_id`, to indicate the type. Then, when the verification system started up, the "program" base class would walk this master list and call the `init()` methods of all the "top" (that is, `type_id==top` objects first, followed by the `init()` method of all the "monitor" objects, in turn followed by the `init()` method of all the "generator" objects, and so on. The actual system had ten different flavors of `type_id`, and thus ten different passes for each method.

Not surprisingly, this "under-the-covers" magic caused significant difficulties for the team. It was hard for subsystems to control which monitors got started in what order, except by carefully controlling which objects were constructed when. Engineers new to the project would get confused and fail to understand the hidden priorities. The team tried to solve the problem by adding a special `StartupClass`, which would be the first to run its `init()` method. However, this made the effort of moving a test from the unit level to the full chip level difficult, because the single `StartupClass` could not be reused. As a result, the "simple" system ended up adding substantially to the complexity of the verification effort.[2]

Example: How apparent simplicity leads to later problems

Here is another horror story. Almost every chip has an interrupt capability. In one case our test team decided to have a single interrupt scoreboard for a chip. The scoreboard would not check the reason for the interrupt; instead, it would simply have a queue of interrupt handlers and make sure there were not any unexpected interrupts or leftover handlers. In practice, this simple scoreboard turned out to be inappropriate for several of the major sources of the interrupts.

There were two main classes in this case. The first was the interrupt handler, which was used to encapsulate the handler logic if it matched the interrupt. This class is shown below.

```
class interrupt_handler:public verification_component {
  public:
    virtual bool match_id(uint32 vector_id);
    virtual void do_handler();
}
```

[2.] One could reasonably argue that this is just an example of a poor or inappropriate design, yet the authors have seen it used in two different companies.

The next class was the interrupt scoreboard. This class had a list of handlers as well as a start() method to watch for interrupts. It also had a post_run() method to make sure there were no unused interrupts. This class is shown below.

```
class interrupt_scoreboard {
  public:
    void post_handler(const interrupt_handler&);
    virtual void start();
    virtual void post_run();
}
```

When an interrupt was asserted, the scoreboard called match_id() for every interrupt_handler on its scoreboard. The first interrupt for which match_id() returned true would be removed from the scoreboard, and its do_handler() would be called. It was up to the test writers to be as specific as they wanted to be in the match_id() method. Some test writers always returned true if the interrupt was for them, whereas other test writers tried to be more specific as to the exact reason for the interrupt.

This class worked fine until the team started testing chip interfaces for which one could not reasonably predict the number of interrupts. Two interfaces in the chip had this property—the USB and the Ethernet subsystems.

In the USB subsystem, a start-of-frame interrupt was generated once every millisecond. Because the USB controller initialized its start-of-frame counter to a random value, and the test end conditions were fuzzy (they were simply based on other data streams draining their checkers), the number of start-of-frame interrupts could not easily be predicted (nor was this number very interesting to know). In this case, the original decision regarding where to put the scoreboard caused an almost impossible checking algorithm for the USB subsystem.

The other example from this same chip was related to the Ethernet unit. The generator randomly assigned masks and packet types, so predicting where (or whether) a packet would arrive was difficult enough—let alone predicting the interrupts that would be generated.

The final straw was that, as an optimization, the chip combined interrupt events, so that if two interrupt-generating events on the same subsystem occurred before these events were serviced, only one interrupt would be generated. Accounting for all these possibilities was not only very hard, but it was also was of little use for verification. As a result, the interrupt scoreboard was removed and individual subsystems were called to handle all interrupts, based on a fixed vector identifier-to-subsystem mapping.

Although a resulting class interface might be too simple for what must be accomplished—and overly simplistic models can lead to complications in implementation—don't stop striving for the simplest usage model possible.

Team dynamics

It may not be obvious, but the makeup and operation of the team affect not only how code is created, but also how well the adaptable code is received. This, in turn, affects the success of the project. Why is this relevant to "thinking OOP?" The addition of OOP created a much more tightly coupled architecture—one where understanding the intent of your fellow team members is essential to coding well. OOP is likely to bring into focus any team issues already present.

A healthy team is better able to create well-built code and adapt existing code. What does this have to do with a handbook on verification? Verification systems have become as complex as production software. As a result, team dynamics becomes a major factor in the success of the verification effort. The sooner we, as an industry, realize this, the sooner we can address team dynamics. *Team dynamics is the current focus of the software domain.* (There are books on this in the For Further Reading section at the end of this chapter.) As this is a new concept for verification teams, we'll just touch on the subject here.

Team roles

There are many team roles and responsibilities. A clear mapping of roles to personnel is necessary for a well-functioning team. An important role is that of the *code leader*. The person in this role is considered the "godfather" of the team and usually is consulted on major and minor architectural decisions. This person knows the language thoroughly and is interested in the latest "best practices." Another important role is that of the *technical leader*, who knows not only the architecture, but also the scripts, policies, and assignments of the team. This person is different from the code leader in that the responsibility of the technical leader is broader, with a more project-level view. The role of *toolsmith* is also critical. This person provides all the scripts and "spells" that make the day-to-day learning and writing of the code easier.

> *Identify and celebrate team roles—they are all equally important to the success of the team.*

Using a "code buddy"

Often an independent reviewer can find places where the code can be made clearer and more adaptable, because the reviewer can concentrate on the finished product, not on the failed attempts. Selecting the right code reviewer is critical to the success of this endeavor.

It is almost always a mistake to have a team code review. This creates a poisonous many-against-one atmosphere. Instead, let each coder pick their personal "code buddy." This individual will be a trusted coworker with whom an informal walk-through of the code can be accomplished. The focus should be on the fact that the code review happened at all, not on the specifics of the review.

> *Code reviews, though necessary, must be done with care. Otherwise, team cohesion and the project will suffer.*

Creating Adaptable Code

How is a section on creating adaptable code relevant to "thinking OOP?" Well, one of the reasons OOP has proved useful is that it really helps create code that is adaptable. In some sense, "thinking OOP" is about creating adaptable and reasonable code.

The term *code adaptability* is an informal measure of how easy it is to move code from one use to another. Code adaptability is an acknowledgment that there is more work to do, even if you purchase verification IP.

Achieving adaptability

Achieving adaptability is a fancy way of saying that you and your programming team creates code that has proved useful for generations of chips. Of course, there is always new code to to write for each chip; otherwise, there is little reason to create another chip. The idea is to build adaptable code that is acceptable in its complexity, can be reasoned about, has minimal assumptions about its environment, and has minimal connections to other code modules, *as appropriate for your organization.*

As an example of code adaptation, one company may prefer to make a copy of some common code before starting a project. In this way they can remove unnecessary code and complexity and, at the corporate level, handle the management activities of common bug fixes. Another company may prefer to keep a single code base for all projects. This will almost definitely increase the code's complexity and size, yet common bug fixes are automatic. Each approach is justified; as is a common theme in this handbook, there are no absolute correct answers.

Let's come up with some of the ways code can be adapted. You can do the following:

- Reuse existing code without any modifications, or
- Copy existing code, or
- Use existing code as base classes, or
- Use only the test cases of existing code, or
- Use only the BFMs of existing code

Here, the meaning of "existing code" includes both in-house and purchased Verification Intellectual Property (IP).

> *The premise of creating adaptable code is that, for the next project or revision, it will be faster to adapt existing code than to develop it anew. This is a major reason for using OOP techniques.*

Why is adaptability tricky?

So most people want their code to live forever. In practice, however, creating adaptable code is difficult.

One tricky thing with adapting code is that the definition of "adaptable" is relative. Does this mean the code can be compiled on different versions of a C++ compiler? Does this mean the code can be reused in a different project? What about verification IP, which can be adapted to different compilers in different projects in different companies? As with the definition of verification, each organization will have different metrics for what makes code "adaptable." The requirements will often change, depending on the team, the individual coder, and the purpose of the code.

While the need to create adaptable code is always present, the actual use of adapted code evolves as both a team and our industry gains experience in reuse techniques. The growing interest in using OOP is a good example of this evolution.

Another difficulty with adapting code is that production code is ugly. A verification system that has made it to tapeout is often riddled with workarounds and undocumented assumptions. It also contains features specific to a particular chip. Moving the code to another project is not trivial.

There are more barriers to adapting code, such as a heterogeneous team experience, a lack of domain experience, and competing requirements for code flexibility.

> *Creating adaptable code may appear difficult at first glance. However, with the techniques presented in this handbook, you can produce code that is adaptable to a large range of projects, with a minimal increase in the code's complexity.*

Architectural Considerations to Maximize Adaptability

This section looks at the reasoning behind, and techniques for, recognizing and building adaptable code. (The next chapter looks more in depth at these considerations). To create an adaptable architecture, begin by asking the following fundamental questions:

- Where and what are the global components (such as a memory map)?

- Where do the lines of responsibility lie?

- What capabilities are needed in the current—and the next—design?

A factor affecting the ease of code adaptation is how close the next chip is to the current one. Obviously, the closer the two chips are, the better the opportunity for adaptation. This touches on the subject of minimizing assumptions in a functional area. The fewer the chip-specific assumptions, the more likely the code can be adapted. However, the resulting code may be more complicated and take more time to develop and learn.

Building adaptable code also involves a *correct by construction* technique (discussed in detail in the next chapter.) This means that, once you change the code for the next design, it is better to have the code fail to compile, rather than having it fall off an "if" test or just return. If the code does not compile, there is obvious work to be done. The worst case is code that compiles and runs, but is wrong. Code assertions, such as the Unix `assert()` macro, can also help to catch the runtime errors. In other words, use assertions for assumptions that cannot be expressed by construction.

Changes are easy—or just plain impossible

An interesting phenomenon occurred when large-scale systems started to be coded in C++. The developers found that changes were either quite easy or nearly impossible. They were easy if the architecture anticipated the change, which usually meant the change occurred along class lines. On the other hand, changes were really, really difficult if several items of data to be changed were spread out into multiple classes. Of course, this is code, and code can always be made to "work," but the very technique of hiding data and making data (objects) drive the algorithms means that some algorithms may be spread out across many objects.

This is relevant because it means that if a feature is not present in the original code, it may be difficult to add that feature later. On the other hand, it might be simple. This leads to a design bias that favors smaller classes, with attention to the assumptions made in those classes. A few examples illustrate this point.

Consider an object that needs to access external RAM as part of its function. If the current design used ZBT[3] RAM, a certain protocol is used. If, in a later design, the RAM were changed to QDR[4] RAM, a new protocol would be used, but the essential function would remain the same. If the original designer had "hidden" the actual memory interface in another object, then it would be relatively simple to build a QDR object and pass that into the other object's constructor.

As another example, consider a testbench object that supports the chip in its initial boot sequence. Suppose that a chip implementation supported booting from UART or USB. In this case you could code the interfaces in the testbench directly, or implement them as specific inherited classes of a generic `boot_source` class. Suppose, also, that later implementations of the chip might support additional boot devices, such as a disk or I^2C interface.[5] The test could then let a testbench choose the boot source. By abstracting the concept of the boot source from the tests, it becomes easier to adapt the tests to new environments and to introduce randomness in testing.

As an architectural bias, making smaller classes with minimal assumptions can lead to code that is more adaptable.

[3.] Zero-based turnaround.

[4.] Quad data rate.

[5.] A serial computer bus invented by Philips that is used to attach low-speed peripherals to a motherboard, embedded system, or cellphone. The name is an acronym for Inter-Integrated Circuit and is pronounced I-squared-C.

Where is adaptation likely to happen?

A fruitful place for adaptation is at the interface, or wire connection, of the chip. This is because an external standard (such as PCI or USB) often defines the behavior there. Note that the programming connection to the interface is not as clear cut as the physical interface. This is where the trade-offs between flexibility and complexity, and the use of minimal assumptions, come into play. (Trade-offs are discussed at length in the chapter on OOP connections; we'll just start thinking about this issue here.)

As an example, monitors on a bus might use events to coordinate with higher-level checkers. Because the monitor does not know whether a checker is waiting on an event, the connection is relatively weak and thus the code might be made more adaptable in this case. As a result, this same monitor class could be used in several different environments, each of which is interested in a different set of events.

Another good area for adaptation is when you exercise a feature of the chip. Conceptually, the code that has to be executed to verify a subsystem is independent of where that subsystem is located. In this case, the trick for maximizing adaptability is to minimize the assumptions the code makes about the testbench environment. If a common framework can be used, including what a testbench provides and the application of a few global objects, it is more likely that tests can be adapted.

Initially, concentrate on the lowest levels of abstraction for creating adaptable code. Then, concentrate on the part of the test that exercises a chip interface or feature.

Separating Interface from Implementation

VHDL encourages a separation of the interface of a module from its implementation. This separation is a bit more difficult in Verilog, which is usually limited to various stub implementations of modules. C++ is more like VHDL, in that it encourages a separation of "what you can do" (interface) from "how it is done" (implementation).

Why is this important? By separating these two concepts, one makes a design not only simpler, but also more adaptable. The design is simpler because the roles of user and implementor are separated. The user of a class is concerned only with its interface and the model of interaction with the class. The implementor, on the other hand, is concerned with how to accomplish the interface. Neither task is simple, but each function can concentrate independently on what it needs to do.

The code is more adaptable because the implementation may change or evolve as the project goes on, or as the code is used in other projects. However, the call that simply uses the class does not have to change. This is particularly important when class inheritance is used, as explained in the next section.

The separation of interface from implementation is an example of the defining of roles and responsibilities of the code. This is a common theme in this handbook, one that is explored in depth in the next chapter.

By separating the interface from the implementation, we make code that is both less complex and more adaptable to different situations.

Interface, Implementation, and Base Classes

Part of "thinking OOP" involves using C++ classes to express common behavior. In fact, expressing one class as a derivative of another is the main mechanism of OOP. But how can this help you write better code? There are two ways: one lets you clearly specify an interface to be implemented, and the other lets you reuse implementations of existing classes. (Don't worry if this issue seems a bit hazy; it is discussed in detail in the chapter on coding OOP.)

To create a class that is to be used as in interface, just use pure virtual methods. For example, to create a top-level test interface, you might have something like this:

```
class test {
  public:
```

```
virtual build() = 0;
virtual void run() = 0;
};
```

Then, specific tests implement the build and run methods, as appropriate for the test at hand. Note the following examples:

```
class dma_test : public test {
  public:
     virtual build() {
     /*code here to build the test...*/
  }
    virtual void run () {
    /*run the classes built by the above method*/
   };
 };
```

The `dma_test` can add other methods as it sees fit; but because it derives from `test`, it must implement the `build()` and `run()` methods.

The other benefit of using inheritance is to save time during implementation. The idea is to have a base class implement the bulk of some common code. Then classes are inherited to implement the specific parts that cannot be generalized.

For example, suppose we had a protocol that can be expressed in a common algorithm, but the final wire-driving mechanism were specific to an implementation of the protocol. Here's how the base class might look:

```
class the_protocol {
  public:
    void do_send(); //implementation in a source file
  protected:
    virtual drive_wire_(bit logical_value) = 0;
}
```

Now, a class can inherit from `the_protocol` and provide the actual driving of the wire.

Another way of using inheritance to reuse code is to specify private inheritance. This is used when you want to use a class, but do not want to bring any of its interface to the users of your class. This technique is often used when "owning" an instance of the other class is clumsy.[6]

Class inheritance is a main part of OOP. Use it to specify an interface, as well as to reuse class implementations.

Summary

• •

We have started "thinking OOP." We talked about how verification is complex, and how this complexity must be managed. We talked about the goal of creating flexible, adaptable code, and how achieving this goal may complicate the code.

We got into coding a bit by talking about the separation between interface and implementation. We also touched on the subject of base classes and the different ways they can be used.

For Further Reading

• •

- The sections "A Complex Solution" and "Accidental Complexity vs. Implementation Complexity" are drawn from the landmark paper, "No Silver Bullet: Essence and Accidents of Software Engineering," by Frederick P. Brooks, Jr.

- The concept of team productivity and its variability is well documented in *Peopleware: Productive Projects and Teams,* 2nd edition, by Tom DeMarco and Timothy Lister.

- One of the "lessons learned" is that the social elements of the team can affect the code you write. Although this is currently an important topic of the software domain, we have decided to stay focused on code techniques. However, there are several good books on the subject of social elements if you are interested.

  The grand-daddy classic is *The Mythical Man-Month*, by Frederick Brooks.

[6.] This often leads to the use of multiple inheritance. A class would inherit using public for the interface, and protected or private for the implementation.

Organizational Patterns of Agile Software Development, by Jim Coplien and Neil Harrison, is a good analysis of several years of software projects.

Lean Software Development, by Mary and Tom Poppendieck, looks at how proven "agile" manufacturing techniques can be applied to software development.

Designing with OOP

*That's right!" shouted Vroomfondel. "We
demand rigidly defined areas of doubt and
uncertainty!"*
The Hitchhiker's Guide to the Galaxy, *by Douglas Adams*

What is design? How do you go about designing with objects? As
far as a CPU is concerned, you don't need objects. You might as well put
all the code into one big function—but that code would be really, really
hard to understand. So we break up a single huge solution into a number
of smaller, understandable, and rigidly defined pieces. This is the essence
of design, regardless of the language.

When we design with OOP, this "breaking up" is really just the construction of a network of classes. However, before we dive into making classes,
we need to understand the OOP design bias. OOP designs tend to focus
on roles and responsibilities. These roles and responsibilities get broken
down further into smaller networks of classes. Ultimately, we wiggle
wires to communicate with the chip.

In this chapter we introduce some basic guidelines for design. We also
talk a bit about some common design mistakes and how to avoid them.

Overview

Design is so intertwined with coding that it's artificial to separate the two. Academic textbooks explain that first you architect, then you design, and then you code. In the real world, however, these steps all get jumbled together. We tend to do all three at once, having a general idea of what we want to do, then refining and changing our idea as we start to code.

Designing with OOP is no different. We talk about "paper napkin" or whiteboard designs. We prototype and refine class interface files and talk about what each class should do. Just as important, we talk about how the classes interact and exchange control and data.

Design is messy, but designing with classes can be a little cleaner. This chapter provides some general guidelines that can help with this inherently untidy process.

Keeping the Abstraction Level Consistent

A key evolution in programming came about when we started to talk about "abstraction levels" in a design. This is somewhat expected, because humans are abstraction machines. A child can recognize a "chair," from the folding chairs at school to the hydraulic ones we use at work, and most people can operate a car, regardless of the make or model. Our mind's ability to abstract away the details of an object or process is directly applicable to programming. We can solve a complex design by using abstractions, from the big-picture operations at the top, down to the wire protocols at the chip interface level.

To put this in fancier terms, at any layer in a design there is an associated *scope of concern* and an appropriate *level of detail*. A scope of concern is the role of the task. The level of detail is the responsibility of the task.

At the top level of the verification system, the scope of concern is the entire chip, its configurations, and the traffic that will be applied to the chip in testing or real life. Here, the level of detail should be very small. In other words, the test should consist of "big" objects, such as the test and testbench, and have no minutia. At the other end of the spectrum, at the bus functional model (BFM) level, the scope of concern should be very small (for example, a handful of pins and wires), but the level of detail should be high (for example, the precise sequencing of those pins to implement the protocol). This is shown in the diagram below:

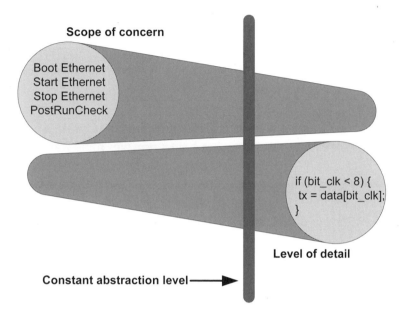

Unfortunately, changes in the abstraction level within an algorithm cause confusion and increase the complexity of the code. For example, shown below (actual code from a coworker) is a top-level algorithm with two shifts in abstraction level. See if you can spot the shifts.

```
void main_process_loop(uint32 num_transfers, IO_BFM*
    io_unit) {
  for(uint32 i(0); i < num_transfers; ++i) {
    BFM_command* command = new command();
    command->randomize();
    io_unit->top->driver->process_command(command);
    for(uint32 j(0); j < 300; ++j) {
      at (posedge(clk));
    }
  }
}
```

The first shift occurs in `io_unit->top->driver`. This is because the engineer trying to understand the algorithm must now understand two more classes. It probably would have been clearer either to provide a `process_command()` method in the `io_unit`, or pass in only the driver to this function. Multiple dereferences on an object are usually a cause for concern.

The other shift is in the at `(posedge(clk))` statement—which is too detailed for the rest of the algorithm. Is it really necessary to worry about clocking at this level?

Object-oriented programming is all about using abstractions! Be sure that a class provides some well-defined service at a fixed level of detail. The implementation of the object is one level down in abstraction, and probably uses lower classes to get its job done, and so on ad infinitum.

Using "Correct by Construction"

As we have mentioned, building a verification system creates a large and complicated network of classes, instances, and conventions. It's not always obvious how to put these building blocks together. However, the C++ language provides strong type checking that can help to communicate the "intent" of the construction. This strong type checking can give clues as to what classes can go together and how they go together. You should strive for systems that, if they can be put together (compiled), are correct. This technique is called *correct by construction*.

Base classes are often used to show intent. A base class can be used to specify a required interface, or to manage a list of homogeneous objects (whose actual types are inherited from the base). For example, a base class called `random`[1] might be used to indicate that a class has randomizable data members, as follows:

```
class random {
```

---

[1] This may also be considered a method convention, rather than a class.

```
  public:
  virtual void randomize() = 0;
};
  class ethernet_settings : public randomizable {
    virtual void randomize();
};
  class pci_transaction : public randomizable {
    virtual void randomize();
};
```

In this case, both the ethernet_settings and pci_transaction can be assumed to have some random data. (Note that in this example, it is probably not appropriate to have a list of ethernet_settings and pci_transaction objects. They are unrelated in function, and related only by inheritance.)

Here is an example that encourages the building of a list of base class objects:

```
  class data;
  class pci_endpoint {
    public:
    typedef enum {in, out} request;
    void handle_data_request(request, const data&) = 0;
    void handle_data_completion() = 0;
  }
  class configuration_endpoint : public pci_endpoint {...}
  class address_endpoint : public pci_endpoint {...}
  class power_management_endpoint : public pci_endpoint
{...}
```

PCI can be viewed as having several types of endpoints, all of which respond to in or out requests. In addition, each endpoint has an associated action after the data have been sent (or received). Thus, it makes sense to have a common base class, and probably a list consisting only of pointers to the base class, pci_endpoint. The actual data in the list will be of the inherited classes.

Enumerations in C++ can also be used to show intent. They can be given names that match the chip's control/status register (CSR) field that they represent and can have values that directly map to the chip, as in the following example:

```
typedef enum window_size {window_4K = 0x100,
                          window_64K = 0x101,
                          window_1M = 0x102};
```

Be careful, because enumerations cannot be extended like classes can. Furthermore, code that uses a `switch` statement on enumerations is a possible sign of design trouble, and the use of enumerations may need to be reconsidered. The warning signs are major amounts of code in the case body, or multiple switch statements on the same enumeration. In this case, an enumeration is being used as control flow mechanism, not as a simple data mapping. It's possible that the enumeration is better represented as a set of classes. Switching on an enumeration is not necessarily wrong, but you must be aware of the code complexity.

C++ also enforces `const` methods and data. This is very useful to show when an object will not be changed (as in `const &`), or when data members will not be modified:

```
class ahb_transactor {
  public:
    uint32 oustanding_transactions() const;
    void print(const string & prefix) const;
  private:
    uint32 current_master_;
    uint32 transactions_remaining_;
};
```

In this case, neither method changes any of the data in the `ahb_transactor`. In addition, the prefix string given to the `print()` method will not be changed.

This is just a quick tour of the concept of "correct by construction." (Many examples of these techniques can be found in the Part IV of this handbook.)

Be careful with systems that must be validated at run time. While some parts of a system will need to use run-time checking, this should be the exception, not the rule. Systems that use run-time checking make it much harder for others to understand your intent.

Base classes, enumerations, and const are good mechanisms for making it easier to see and enforce how a system can be put together.

The Value of Namespaces

When a large amount of code is developed, there will be classes or functions that have the same name. If you had access to the source code, you could standardize the colliding names. However, changing the code is more difficult when the code is from another work group, division, or company. C++ helps to minimize this problem by providing *namespaces*. A namespace is like a class, in that the methods (and data) in them must be accessed with the name of the namespace. By using namespaces, you minimize the number of global names, because the previously defined global functions are all in namespaces, and now only the namespace identifiers themselves are global. This decreases the probability of a collision.

Namespaces are useful for grouping related classes. In the UART example chapter later in the handbook, there are several classes for the generator, BFM, and configuration. All of these classes are wrapped entirely in a `uart` namespace. A good design bias is to have each of the verification components in your system be in their own namespace.

For example, in a file called `lcd_display.h`, you might have something like this:

```
namespace lcd_display {
  class generator {...};
  class checker {...}
  void start(); //start the generator and checker
  void stop();
}
```

You can use a namespace by specifically naming it or making it implicit by means of the `using namespace` keywords. Below is an example.

```
#include "lcd_display.h"
using namespace lcd_display;
//now lcd_display functions may be used without
    qualification
void lcd_function() {
  lcd_display::generator my_generator();
  lcd_display::start();
  checker my_checker(); //resolves to
    lcd_display::checker
  stop();                 //resolves to lcd_display::stop()
}
```

Be extremely cautious of putting a namespace using clause in a header file. It is almost always a mistake, because every file that includes your header file will inherit the using clause's scope. In addition, every file that includes the header file that includes your header file will now have the using clause—and so on, with possibly unintended consequences. The authors have first-hand experience of trying to undo this technique. The task was not pretty.

Another use for a namespace is to wrap a related set of global objects and functions. The interface is just a collection of functions, but the wrapping of them into a namespace creates what is called a *singleton*. While you want to minimize the number of singletons in a system, they are necessary and correct for those areas that represent global resources. Singletons are discussed more in the Coding OOP chapter.

One last point about namespaces. You can use an anonymous namespace within a source file instead of marking your local utility functions static. The examples in Part III of this handbook use this technique.

Namespaces are very useful for grouping related classes and functions. Be careful with the using *clause in header files.*

Data Duplication—A Necessary Evil

There is an inherent trade-off between minimizing the complexity of connections among verification components and duplicating the data passed among them. The looser the connection, the more likely that there is a duplication of data. This is talked about in detail in the next chapter, but we'll discuss it as a design bias here.

Let's look at an example of duplicated data. Consider a DMA chip's view of data versus a checker's view. A processor sends to the DMA engine the source address, the destination address, and the length of th e transfer. When the DMA completes, selectged memory contents are stored in an object that contains a start address, an end address, a length, and a completion status for the transaction. The chip synchronization mechanism for the monitor is probably an interrupt. The class and the chip memory both contain the data, so we have a data duplication situation. In this design the data duplication is good, because checkers can use the more abstract fields of a class, instead of reading memory directly. Also, the code is more portable, because the concepts of source, destination, and length are abstracted away from the actual memory layout.

In the previous example, the data representation changed from chip memory to fields in a class. However, there is another place where data are duplicated. This is a more subtle, yet useful, design technique.

We think nothing of calling a function and passing it an integer. We know that the integer's value will be copied. But what if this integer were the result of some computations in the function making the call? Now the situation is not so clear. Why? Because this integer is not just a value. It represents the result of an algorithm that computed some data. The fancy term for what this integer represents is *derived data*. This is a normal situation, and it is used all the time when abstraction layers are crossed.

In fact, derived data is common in multilayer protocols. Many protocols have a physical layer, which deals in bits and words. Then they have a transport layer, which deals in packets. Some protocols even have a third layer, which deals in higher-layer transfers. As each layer hands off to the next, data are copied. The concept of derived data also exists when abstraction layers are used in verification.

For example, consider the physical interface for Ethernet Media Access Control) MAC. Suppose a verification component called MAC supports both Media Independent Interface (MII) and Reduced Media Independent Interface (RMII), and expresses this in an enumeration. These two interfaces are quite close, with RMII being a reduced set of the MII interface. In order to simplify the control connections between the MAC and the physical layer, the connection could be a single bit set to one for "full," to distinguish it from "reduced," where the bit would be zero. Although the information in a single bit is weaker and simpler than the passing of an enumeration, the bit is more appropriate to the lower-level abstraction.

Be aware of the same data being in two places at once. This generally happens across abstraction levels. Be very aware when derived data must be refreshed.

Designing Well, Optimizing Only When Necessary

When hardware is designed and implemented, efficiency is critical to its ability to be built (meeting timing, power, and size constraints). However, in software the emphasis is often on minimizing mental complexity. This is natural, as increasingly complex control and data structures can be built relatively quickly with software.

In verification, you must focus on clarity in your overall design and implementation. This does not mean, however, that you can ignore efficiency.

The speed of a functional simulation almost always depends almost always on just a small percentage of the actual code. You should optimize the code for simulation speed only after you have finished profiling it thoroughly. Premature code optimization leads to confusion. While it might run a simulation a little faster, it will more often than not slow down the project by being harder to understand and use.

Get the design working and understandable first. Then figure out what needs to be optimized—and possibly made more complicated.

Using the Interface, Only the Interface

The verification test system must both apply stimuli to the chip and then check to make sure the chip generated the correct response. Consequently, the verification system is aware of *both ends* of a data path—an awareness the software does not have. In a simulation, creating both ends of a data path is necessary, but it can lead to sloppy code that may mask chip problems. Be aware of what information the chip interface provides, and use only that information for checking.

For example, the authors helped test a USB device. The original verification tests preloaded the chip with the "correct" data. Then, when the data were needed, as specified by the protocol, they were sent by the chip to the checker. The checker confirmed the data and all was well. Upon closer inspection by the software team, a critical parameter was found to be missing: the length of the response. By design, this parameter was being dropped by the chip, but because the verification code already knew the "right" answer and length, the design error was missed. Had the verification code been cleanly separated into a *generator* (the requestor) and a *checker* (the interpreter) of the request, the error would have been caught.

Another example concerned a DMA checker. The DMA checker required the verification components to register all the memory accesses that the chip would make. At the end of the test, the DMA checker would make sure there were no unmatched writes. What the checker failed to test, however, was the result of overlapped memory writes. There was a bug in the cache coherency unit that was masked because the tests did not write to the same address twice.

> *Minimize the assumptions made across different parts of the verification system. Strive to use only the "knowledge" that the production software has through the feature or interface.*

Verification Close to the Programming Model

Verification is both similar to and different from the production software that will run on the system. However, the closer the verification architecture is to the software, the better the chance you have of finding errors in the programming model, with respect to condition status registers (CSRs), hardware algorithms, and so on.

So while it is true that the software you use for verification is inherently a lot more detailed than the production software, you should still strive to keep your test algorithms as close to the production ones as possible. This design bias not only helps the software and hardware folks communicate better, but it can also help shake out register/memory/interrupt design and programming model.

For example, the authors had a project where the design of the DMA part of a chip was changed after the verification team had their code reviewed by the software team. Yes, this change was humbling. Specifically, the chip used a bit to restart a DMA channel. The original design had made the transition of zero to one be the enabling action. This meant that the software would first have to set this bit to zero (it was a one from a previous enable!), and then set it back to one. Once the software team was aware of this issue, they asked how hard it would be just to restart with a write of one, even if the previous value were a one. This was done, making the DMA enable more intuitive to the software team. Another change concerned the ability to write the DMA offset and index. The original design allowed only a reset of the index to zero, which made resending a DMA buffer clumsy. The software had to recopy all the buffer descriptors in memory to the zero offset address. The hardware was changed to allow the index to be written whenever the DMA unit was paused.

A final change was in the address map. Previously, the address space for the DMA was contiguous, which made sense to the hardware team. However, because the various I/O subsystems were features that end users paid for individually, the software had divided up the device according I/O subsystems. Because of this division, there were two different address spaces to manage: one for the I/O subsystem, and another for the DMA. We changed this design and moved the specific

DMA channel's CSR address space to the I/O subsystem that it served. While this change seems (and probably is) simple, it shows that the hardware and software teams viewed the chip differently. Working more closely together improves the overall quality and helps reduce total project time. Furthermore, using C++ allowed the two teams to look at each other's code without too much complaining.

> *Try to design your verification system to use the same algorithms and subarchitecture as the production software. That way, you can catch clumsy or conflicting programming models.*

The Three Parts of Checking

The major thrust of verification is checking the operation of the chip. We send data in, or turn on features, and check that the chip produces the correct response. A fancy term for the transmission of data and the enabling of features is called *stimulus*. As a design bias, it's a good idea to separate the generation of the stimulus from the acting on and checking of the stimulus. We'll look at this further in subsequent chapters, but for now let's look at the *checker*.

Once data have been injected into the chip, the checker recovers the actual data output from the chip's I/O and confirms that everything is as expected. There are three parts to this process and, to promote adaptability, the implementations of each part should be kept separate. The success of code adaptation can often be traced to how well these three parts, summarized below, can be shaped to fit a new environment.

- The first part is the gathering of the data, usually by means of a monitor. A monitor triggers on changes in some I/O, interrupt, or FIFO level and converts these wire changes into verification objects. By having a monitor that is separate from the checker, you can change how the data are gathered without affecting the checker. Also, by converting the data from wire changes to integers and classes, you elevate the data by an additional level of abstraction.

- The second part is a comparison of the actual data with the generated data. This can be accomplished by providing a global `operator==()` method. (Part IV of this handbook shows examples.) This is useful because you can change either the data or the comparison, yet the checker remains the same.

- The third part of checking uses the result of comparison and provides an indication of expected or erroneous behavior. The simplest example is an error message or "check passed" message, followed by a printout of the data.

In the checker class, be aware that the form of the data generated may not be the same as the form received by the monitor. This is because data packets may have been combined or split for a variety of regions (for example, because of the protocol, error correction, or some other transformation). Creating an appropriate level of abstraction for a checker can be difficult.

Sometimes there is more than one level of checker. This is common in multilayer protocols, such as PCI Express, Ethernet, ATM, and USB. Again, keeping the levels separate improves the code.

Sometimes there may be several monitors on the same chip I/O. This is common because in the verification code a one-to-one relationship between monitor and checker is the simplest, while in the HDL there are no simulation limits on the number of monitors on a wire.

The checker should also check for dropped or missing data. The easiest way to do this is at the end of the test, but it's probably better to use the latency of the chip as a filter, and report errors as soon as possible.

Checking consists of three parts: gathering data, making a comparison, and acting on the comparison. Separating these three parts is a "flexibility vs. complexity" trade-off.

Separating the Test from the Testbench

It is common to have a few top-level parts to a verification system. These include the HDL testbench, the C++ testbench, and the test. The authors advocate for another top-level class, the verification top (see the Layered Verification Approach chapter earlier on). This section addresses the roles and responsibilities of each part.

A major theme of this chapter is to show how clear roles and responsibilities create simpler code. In general, for a single chip try to have a single caller, a single C++ testbench, one or a few HDL testbenches, and many tests.

The verification top is the top-most level of a test. The method calls are the same for every test, and represent the standard steps of creating, configuring, running, and shutting down the test. Most projects already have a standard verification top.

The C++ testbench is responsible for setting up the transactors, monitors, and generators (under direction by the test), and building the global services. The HDL testbench is responsible for the HDL wrappers around the chip, and includes clocks, reset logic, and pin wires. In addition, the HDL testbench probably contains muxes, assigns, or Verilog `tranif`[2] statements for connecting verification wires to the chip. It may also have wires for "power on" (to communicate with the boot I/O devices and enable the initial state configuration of the chip). The test is responsible for specifying the required verification components, traffic patterns, and verification configurations, as well as what the "run" part of the test should do.

Having a single C++ testbench for a specific chip or system is useful, because it sets up a common environment for all tests to use. This increases the adaptability of the individual tests and verification components. Also, because unit testing differs from full system testing, it may be necessary to have a few HDL testbenches; however, effort must be taken to ensure that as many of the tests as possible can be run in either environment.

[2.] The Verilog primitive `tranif` connects two bidirectional wires.

Because a test must specify what configuration it requires, there is some communication between the test and testbench, possibly before the chip can be brought out of reset. This means that, while the verification top "news" the test, the test may make several calls to the global objects, possibly including a verification components manager, to configure the testbench.

As an implementation detail, the test may be "newed" by a *factory function*. (Factory functions are explained further in the Coding OOP chapter.) This function, usually implemented in the source file that contains the test, returns a base class test pointer. The reason for using a factory function is to incorporate unit-level tests into a full chip test.

Sometimes this base test object contains all the verification components and a basic structure for the test. This can be useful, but be aware of all the types of tests, and don't make the base test too complex.

A test can have a few standard components: the dance, the C++ testbench, the HDL testbench, and the test interface. Only the implementation of the test should change.

Summary

We have started down the path of using OOP in a verification system. We talked about the main theme, creating roles and responsibilities by using abstraction. We talked about the common design biases used when we design a verification system.

You probably are still surrounded by clouds of uncertainty. This is understandable. The next chapters are more specific, talking about making classes and the different ways to connect them.

For now, however, know that designing with OOP is about defining roles and responsibilities and making levels of abstraction, a "layering" for which there are many examples in our everyday lives. To achieve your own design objectives in silicon, use your experience to guide the process.

For Further Reading

- To help you think about how to construct a system with abstraction levels that are logically consistent, a great book is *The Design of Everyday Things,* by Donald A. Norman. Though it does not deal with code or high-tech, it is great for thinking about how someone else might develop a mental model of using your code.

- On the topic of connections between levels of abstraction and within a level of abstraction, *Software Engineering: A Practitioner's Approach,* by Roger S. Pressman, has several pertinent sections. (The fancy term for these connections is "cohesion and coupling.")

- Bjarne Stroustrup also provides a concise discussion of abstraction in *The C++ Programming Language*, section 24.3.1: *"What do classes represent?"*

- The concept of "correct by construction" is from Edsger Dijkstra, a pioneer in formal languages, specifications, and proofs for computing. This concept is often used in formal verification, but it is adapted here to show how design intent can be communicated with C++.

- Regarding premature optimization, the original quote is from Tony Hoare: *"We should forget about small efficiencies, say about 97% of the time: premature optimization is the root of all evil."* This was quoted in Donald E. Knuth, *Literate Programming* (Stanford, California: Center for the Study of Language and Information, 1992), 276. A web search can provide further references.

OOP Classes

Experience is a dear teacher, but fools will learn at no other.

Benjamin Franklin

Coming up with the appropriate classes for your project is an experience-based effort. In other words, the authors made many mistakes in the beginning. To help you in designing classes, we have collected experience from our previous efforts.

So do what the authors did when they learned C++: find examples, copy, and paste!

This chapter introduces the thought process for creating classes, to answer questions such as these:

- How do I determine what is a class, and what is a method?

- How should I handle global functionality?

- What can inheritance do for me?

- What can operator overloading do for me?

- What does the C++ compiler create automatically?

Overview

Classes are fundamental to writing in an object-oriented language. But how do we decide what is a class? We have talked about thinking in terms of layers of abstractions. We have talked about roles and responsibilities. The next thing is to start to name the classes and their responsibilities. This is not as hard as it sounds. For one thing, you make classes as you feel they should be, and there is no right or wrong way. Let each class do what feels right to you. There will, of course, be some spirited team discussions. This is the topic of the first section of this chapter.

Once you decide on some classes, you can "wire up" instances of classes pretty much like you create and "wire up" modules in hardware. Unlike hardware, however, classes can have more "electricity." When designing hardware, you are restricted to connecting blocks through wires or signals, but with classes in C++, you have the ability, among other things, to have pointers to other class instances or call virtual methods. This is the topic of several sections.

This additional freedom is where the electricity comes in. This is good, because it helps you solve complicated verification problems. As with any technique, the challenge is to use the appropriate amount of electricity.

Not that everything has to be a class. C++ supports good old global scope functions, and for many situations, functions are appropriate. The section in this chapter on Global Services talks about various ways to package global functions in C++.

We end the chapter with a quick tour of some advanced class features of C++. We don't do this to make you concerned; rather, the purpose is to let you know that the language can solve pretty much any problem you will encounter.

Defining Classes

As C++ has been around for many years, there have been many different attempts to explain how to define good classes. In the end, it comes down to the usual way one learns: copy examples, change the example a bit, and eventually start writing your own code. After some time you will find your own way to "ride the OOP bicycle." That's the reason this handbook presents lots of code snippets and examples.

That said, a common way to define classes basically just follows the old grammar school rules for writing a good sentence: make a class for each noun in your design, and make a method for each verb. This means that each block in your whiteboard design becomes one or more classes. Drawing the lines between the blocks is a bit more tricky. At some level these lines represent method calls, but they can also be classes themselves. That's the great thing about C++ compared with HDL design; you can use a variety of alternatives (more language constructs, more techniques allowed by basic constructs) as you discover problems in implementing the initial "obvious" whiteboard class design.

It is promising that the industry has finally settled on (more or less) standard names for the most common classes. Names such as *generator*, *BFM*, *monitor*, *driver*, and *checker* have become somewhat standard in their meaning.

> *Making classes becomes only easier with experience. First clone and modify existing code.*

How Much Electricity?

C++ is a language that allows an enormous amount of "electricity" in the code. The electricity in a piece of code is a measure of how complex the code is. Recall that complexity is inherent in our world; it's the management of complexity that prevents complex code from becoming complicated code (remember that complex is okay, complicated is bad).

The goal of any OOP system is to design classes that have minimal electricity. Anything more is just unnecessarily complicated.

At the lowest level of electricity are *defines*. In order to figure out the code, you have to figure out the value of the define. The next level is "if" tests, which are dealt with in the chapter on Coding OOP. Verilog has these capabilities as well.

Classes

The next addition of electricity is *classes*. Verilog has the *module* concept, which is pretty close. Both concepts unite data and algorithms (*tasks* in Verilog, *methods* in C++); they differ in that in C++ the data can be classes, whereas in Verilog the data are wires and registers. Modules in Verilog represent silicon, whereas in C++, classes represent a wide range of concepts.

Namespaces

Related to the concept of classes in C++ is *namespaces*. This is somewhat similar to the *package* concepts of VHDL and SystemVerilog. Namespaces are useful in that they group related classes, global functions, and data together loosely.

Operator overloading

Where OOP languages differ in capabilities from Verilog starts with operator overloading. This capability allows your classes (and `enums`) to act like built-in types, which is necessary for algorithms to have minimal complications.

For example, you could make a register class look like a built-in type. Using operator overloading, you can make your register class add, subtract, shift, assign, and compare just like a built-in type (that is, by using operators such as +, -, >>, = and ==). In addition, you can add Verilog-specific concepts, such as slicing and printing in binary format. Operator overloading is also used for the C++ standard library string class.

It's worth noting that operator overloading in strings is not a special feature of C++, as it is in Vera, SystemVerilog, and "e." The standard string functions are just normal C++ code that makes the string concept look built in. Because overloading is a standard feature of C++, anyone can add new classes, as needed, that act like built-in types.

Pointers and virtual functions

Another increase in electricity relative to operator overloading is found in *pointers* and *virtual functions*. This relatively simple addition to C has profound implications. The number of techniques that can be realized with this electricity has spawned numerous books and papers. Remember that at the implementation level, virtual functions are pretty much just plain old pointers to a C function. You may have used this technique before.

Even more electricity

C++ even provides other kinds of electricity, such as templating, method overloading, and class level `new()`, `delete()`, `operator->()`, and `operator()`. While we may use some of these techniques, they are not normally part of the main code of verification systems.

Smart pointers

Before we leave the discussion of electricity, we should talk about *smart pointers*. In some verification languages, such as SystemVerilog, Vera, and "e", these pointers are the only class-referencing mechanism. This is unfortunate, because, as with operator overloading, they could have been added as a generic feature, to be used, or not, by you.

Because smart pointers manage the pointers to a class, counting the number of references to it, by watching the `operator=()` and the copy constructor, they simplify the management of memory. In general, they make the assumption that everyone wants to share the same data. However, this is obviously not the case universally, depending on several factors. So, instead of letting the *single programmer* of a class decide (as is done with standard strings, for example), the *many users* of the class must explicitly call a copy or "deep copy" method to break the connection of shared data. This is clumsy and creates unnecessary complications in the user code.

C++ provides many levels of electricity. Use minimal electricity in your designs.

Global Services

OOP-based design is about using abstractions and defining roles and responsibilities for specific classes. By using layers, as described in the Layered Approach chapter in Part I, you can simplify the design and set up a network of classes. For example, a monitor and a checker have a *neighbor* relationship; the monitor takes data from the chip and the checker checks the data.

However, some roles and responsibilities are related to a large number of other classes. Activities such as memory reads and writes, control and status register (CSR) reads and writes, interrupt vector handling, and message logging can reasonably be expected to be available to all classes. There are many more examples. Roles and responsibilities that are available to all classes are called *global services*.

A logical way to create global services would be to use a class for each service. Then, pointers to these global service classes could be passed to all other classes. However, in practice this is often a clumsy approach. Passing the global service objects to the majority of all classes clouds the code and adds little to the real information of a design. Used with restraint, though, global service objects can extract the common components from the mental baggage of learning a new design structure.

Namespaces

The authors prefer to use a *namespace,* or sometimes static methods of a class, to express the intent implicit in a global service. As a result, any class can include the header file and then use the service.

For example, consider a memory namespace such as the following:

```
//in the file memory.h ...
namespace memory {
  void write(uint32 address, uint32 data);
  uint32 read(uint32 address);
}
```

Note that read() and write() are memory functions. For this service, any class can include memory.h and start accessing memory through these functions; no special rules or objects are needed. Each class accessing the memory functions still needs to use the memory namespace, which gives the reader a way of finding the source of these functions.

Therefore, using namespaces can simplify a verification system while keeping the source available in a central place. In addition, any class can access memory without having to know about how the memory is implemented.

A couple of other nice features about namespaces are worth mentioning. First, you can continue adding to a namespace by simply *redeclaring* it. At a later point, when you are ready to add some new code to an existing namespace, you can simply put the same namespace tag on the new code.

For example, to add to the memory namespace we talked about above, you would say the following in any other header[1] file:

```
namespace memory {
  class write_burst(uint32 address, std::deque<uint32>
    data);
  //...other code, as needed
}
```

Now, by including both the old memory header file and this new one declared here in your source code, the memory namespace contains not only a write and a read, but also a write_burst function.

The other point to mention is that there is a "default" *unnamed* namespace that is created for each file. This is for the global variables you need for that specific file. These global variables can reside in an unnamed namespace, so they don't collide with other global variables for other classes. You declare an unnamed namespace like this:

---

[1] A header file is, by C++ convention, a file that just provides the interface of a class.

```
namespace {
  static uint32 counter(0);
}
```

You access it like this:

```
if (::counter > MAX_CNT) ...
```

Note the ":::" before the counter name. This is not required, but it does help to show that this is a global. Using the global namespace is a good way to "hide" variables that need to be global only to a file. Otherwise, if another file also has a global variable of that same name, the linker will—sometimes silently—choose one.

Static methods

Another way of presenting a global service is to use *static methods*. Using the same memory example as above, you could instead declare the global services in a class like this:

```
class memory {
  public:
    static write(uint32 address, uint32 data);
    static uint32 read(uint32 address);
};
```

The way to access the global service in this example is to instantiate a memory object in the code. The end results are similar to those achieved by using namespaces.

Singletons—A Special Case of Static Methods

Instead of using all static methods in a class, you can use a single method to get the one instance. After that single method is used to get a pointer to the single instance, the accesses show up as with any other object. This can be useful if you want to communicate that there is a class instance performing the work. On the implementation side, it can make the different implementations of the (previously declared) static methods easier. At any rate, using this technique implies that there is more electricity than is the case with just a namespace or static methods. Let's look at the memory example again.

```
class memory {
  public:
    static memory& get();
    virtual void write(uint32 address, uint32 data) = 0;
    virtual uint32 read(unit32 address) = 0;
}
```

The single static method, get(), is used to get the single instance of the class. This technique is called a *singleton*. Notice that the read and write methods are virtual. Let's look at how this might be useful in a burst memory class.

```
class burst_memory {
  public:
    memory& get();
    virtual void burst_write(uint32 first_address,
                             std::vector<uint32> data);
    virtual std::vector<uint32>
                             read(uint32 first_address);
    virtual void write(uint32 address, uint32 data);
    virtual uint32 read(uint32 address);
}
```

The implementations of the read and write methods are probably simpler now that a singleton is used. Otherwise, the read and write methods would both have to deal with gaining access to the correct memory.

Namespaces or static methods?

So why prefer one technique—namespaces or static methods—over the other? The reason is that namespaces are more expandable than static methods, because they are less complex. In addition, namespaces can easily be added to, providing a simple way to extend the global service.

With static methods in a class, you run into trouble when you need to add a new static method. Doing this requires that you inherit from the original class and add the new static method. Although using static methods for global services is useful for more component-like services, using a namespace is often easier than using a static class method. Nevertheless, despite this advantage of simplicity, there are times when you want to bring to the interface the fact that the implementation is an object. That is when singletons may be appropriate.

Using namespaces is a good way to provide access to global services, and it is less complicated than using static methods. Although singletons are a good way to implement a global service in a class interface, be aware of where the object is created.

Other considerations

One last point before we move on. A global service, at the interface level, implies a logical single service. The actual implementation of this service, however, may be vastly different behind the scene.

For example, to write to different memory addresses, several different objects may be needed. This is because the memory space is probably spread out across different chips—or at least across different interfaces of a chip. Also, if you fold the register access into the memory access, different register banks may communicate with different chips. Finally, you may want front- and back-door access for the memory and registers, which would require different objects to do the implementations. This can all be an implementation detail for the end user.

Another technique is to implement a singleton as a list of filters. This is useful for purposes such as filtering logging messages. For example, a logger might look to the classes that use it as a single object. The implementation, however, may want to use a linked list of objects, where each object is given the chance to modify the log message. This is a powerful technique. (To see an example of this, refer to the implementation of the `vlog` class in the `vout.cpp` file of Teal on the accompanying CD.)

Often there are global services in a verification system. Use either namespaces or static methods to express them.

Class Instance Identifiers

When you start printing log messages, you have to decide how to identify the object that is printing. This object identification provides a way to trace the object through the system.

There are at least three techniques for identifying an object. One uses the address of the object; a second uses a string (often the name of the instance); and a third uses some sort of sequence counter. In practice, all three techniques, as well as their combinations, are used.

Object memory address as identifiers

One of the benefits of C++ over other HVLs is that C++ has a pointer concept. The pointer is a unique address to an object. This can be useful as a unique ID for tracking an object. However, this is generally appropriate only for utility objects, because printing the address means little to a user.

Strings as identifiers

As you move up in abstraction level, it will at some point be better to have names for objects. These are most often placed in the constructor of a class, as follows:

```
class fabric {
  public:
    fabric(const std::string& name);
    virtual vout& operator<< (vout& v) {
      v << name_; //and other parameters
    }
  private:
    const std::string name_;
  }
}
```

With the class using a string for a name, it should be easier to print useful status messages. Because the name is passed in, it is easier to make a unique name reflect its use in the chip.

Static integers as identifiers

Sometimes it's too cumbersome to use a string as an identifier for an object. Also, the pointer, while unique, does not indicate a sense of sequence between consecutive objects.

For example, often a generator creates a sequence of objects, as in, say, a number of Ethernet packets. In this case, it may be appropriate to name the instances with an incrementing integer, such as `packet_1`, `packet_2`, and so on.

Having sequence numbers can be useful as a triggering mechanism for trace types of logs, or for postprocessing the log file. This is done in C++ by making the counting integer `static`; this declares a class level "shared" integer, and increments it for each instance of a class.

Here is an example:

```
class data_packet {
  public:
    data_packet() : my_id_(id_++) {}
  private:
    uint32 my_id_; // count of data_packets created,
                   //starting with 0
    static uint32 id_;
}
```

Then, in the `data_packet.cpp` file, you would define the storage for the single class-level integer, as follows:

```
uint32 data_packet::id_(0);
```

Combination identifiers

In practice a combination of these techniques is often used. For example, one may want to prefix the sequence number with a name, which can identify the higher-level sequence (such as "short packet #43," or "Device 7: packet 10 Enumeration Phase").

Identifying an instance should not be an afterthought. Often, a pointer address is sufficient. For sequential instances, a static counter can be useful in tracing an instance in the log file. For most classes, an instance name should be included.

Class Inheritance for Reuse

When you start to name and build classes, there is a tendency to find commonalities in roles and responsibilities. While this is certainly a good thing, resist the urge to define base classes right away. There are many ways to express common roles and responsibilities. Use base classes only when you have experience from several designs, or when you are actually coding and can use base classes to solve a problem. With these caveats in mind, let's look at inheritance and how it can help to make code more adaptable.

In verification we have to drive and monitor the wires of the chip to exercise an interface with what is commonly called a bus functional model, or BFM. One side of the BFM is connected to the data generators and monitors; the other side is responsible for driving and monitoring the wires of a chip's interface.

It is good practice to separate the actual driver or monitor into two separate interfaces. One is the data generator or monitor interface, and the other is the BFM interface. This is an inherent separation point, because there is a conversion between abstraction layers from "send packet #3" to the actual wiggling of the wires. This separation minimizes the assumptions about how the BFM does its job.

This separation is a good use of inheritance.[2] The authors call the base class that wiggles the wires the *BFM,* and the inherited class that deals with interfacing with the generators and monitors the *BFM agent.*

A BFM base class example

Consider a chip interface for sending and receiving packets. The base class might look something like this:

```
class packet;

class bfm {
  public:
```

---

2. There are other techniques for this natural separation. You also might just want to own the BFM or use private inheritance.

```
bfm(/*some connection to the wires*/);
virtual void start(); //start the receive thread
    running
//This is the driver part of the bfm...
void send_packet(packet*);
//This is the monitor part of the bfm...
virtual void packet_received(packet*) =0;
}
```

The `packet_received()` is declared as a pure virtual method, so it must be implemented by the inheriting class. This is because the BFM should not know what to do with a completed packet, but just focus on how to drive the data.

A BFM agent class

Now an inherited class can add channels (one for "received" and one for "send") to convert from packets to commands, and vice versa. The inherited class would look something like this:

```
class bfm_agent : public bfm {
  public:
    bfm _agent(/*two channels, and the wire interface*/);
    virtual void start(); //start BFM and then thread
                          //to get data from the send queue
    virtual void process_send_command();
    virtual void packet_received(packet*);
}
```

Reusing the BFM class

We could have just lumped the `bfm_agent` and `bfm` into one class. However, with two classes the responsibility of each is better defined, although things can become a bit more complex.

When another project comes along with a chip that has the same BFM interface, but that accepts only a small fixed set of packets, the new test team can still use the `bfm` class and write a simpler `bfm_agent`. Also, because the new `bfm_agent` class is so simple, they might decide to include the checker directly as part of the agent class, and not use a channel at all.

If the team so decides, their project can still reuse all of the existing bfm class code by just inheriting from the BFM with a new, simpler, combined generator/checker class. This is why dividing models into layers of classes is a good idea.

Using inheritance to adapt code can preserve working code and still add features.

Class Inheritance for Interfaces

The previous idiom of using inheritance for reuse is common. We can take that to the extreme, and instead of reusing the implementation, reuse the interface only. This is more useful than it sounds. In fact, this technique is very powerful. You can use the interface of a class to communicate exactly what the classes in the hierarchy can and cannot do. This helps a reader of the code build up a "mental model" of the system. You can only call the methods defined in the base class—no more, no less.

For example, as shown in the Part II of this handbook, a common design for a testbench is to have a top-level procedure that builds a number of high-level, independent verification components (usually related to the major interfaces or features of the chip).

In order to manage the complexity of the verification system, it is important to have common base classes for these major components, so that all the components in a verification system behave in a similar, predictable way. If each major component is built in a unique way, it soon becomes too difficult to manage the overall environment.

Inheritance for a verification component

Let's look at the following verification_component class:

```
class verification_component {
  public:
    virtual void initialize() = 0;
    virtual void start() = 0; //forks threads
    virtual void stop() = 0; //stop and join threads
};
```

Now lots of common verification components can express themselves as an inherited class, like so:

```
class test : public verification_component;
class BFM : public verification_component;
class monitor : public verification_component;}
```

By having all verification components inherit from the verification_component class, one can understand that each class can at least be initialized, started, and stopped, because these are pure virtual methods that exist in the base class.

Inheritance for a payload interface

Because using inheritance for interfaces is so common, let's look at another example. In verification there is often a payload of data that travels through the system. The data must be random, printed out at various times, and compared with the initial data sent in. If the chip had an Ethernet interface, for example, there could be an inherited class that extends the base payload data class into an Ethernet payload class.

Here is what the base payload class might look like:

```
class payload_base {
  public:
    virtual void print(const std::string prefix) const
      = 0;
    virtual void randomize() = 0;
};
```

Note that the payload_base class has no data; because this a generic base class with no data, there is no constructor.

The base class declares two virtual methods, one intended for printing and another intended for randomizing the payload of any concrete class. For an Ethernet packet the base class could be used like this:

```
class ethernet_data : public payload_base {
  public:
    virtual void print(const std::string prefix) const{
      for (int i(0); i < data_length_; ++i) {
        teal::vout("ethernet") << prefix << "data[" << i
                    << "]" << data_[i] << teal::endm;
      }
    }
    virtual void randomize() {
      std::for_each(data_,RAND_8);
    };
    virtual bool operator==
                     (const ethernet_data& other_one) {
      return(this == &other_one) ||
           ((other_one.data_length_ == data_length_) &&
            (std::equal(data_.begin(), data_.end(),
                         other_one.begin())));
    }
    void put_to_DUT();
  private:
    std::vector <unsigned char> data_;
};
```

Shown above is an `ethernet_data` class that inherited from the `payload_base` class. As required, this new class provides specific implementations of the `print()` and `randomize()` methods. It also adds a new method, `operator==()`, as well as a `put_to_DUT()` method.

The `operator==()` method is interesting in that it will be called automatically when two objects of `payload_base` are present in an equality expression, that is, "`if (payload_base_1 == payload_base_2)...`" Note that this is not magic; the compiler keeps tables of the methods, and calls your methods instead of generating a comparison of data member by data member.

The `ethernet_data` class also adds the `put_to_DUT()` method, which is intended to transfer the payload to the chip.

These four methods—`print()`, `randomize()`, `operator==()`, and `put_to_DUT()`—are the *only* things you can do with an instance of an `ethernet_data` class.

Note that you cannot restrict the use of a base class by inheritance.[3] By the definition of C++, anywhere a base class is used, an inherited class can also be used.

Inheritance can be used to communicate exactly what an interface can do and or must implement.

Operator Overloading for Common-Currency Classes

With C++, classes and enumerations can act just like built-in types. This is essential when you want define a class or `enum` that a user can use intuitively in binary operations (such as <, ==, !=, +, and *) and unary operations (such as +, ->, and !). You can also provide specific behavior for operator `new()`, operator `delete()`, `operator->()`, and `operator()`, which provide a wide range of powerful capabilities. The C++ template library has some good examples of this.

However, be careful about making complicated objects use operator overloading. This is because the more complicated an object is, the more difficult it is for it to act like a simple built-in type. What happens is that there will be two or more completely valid definitions for one or more operations. Consequently, it is better to stick with small, simple objects that act like built-in types. A possible exception is `operator==()` for objects that will not be inherited from. In this case it may be cleaner to put the comparison code into the object.

Note that there is a large number of C++ operators. Which ones you provide depends on the operations and interface you want to support. It is not a simple task to determine the "right" subset, but start small and add operators as needed.

[3.] Except, of course, in protected or private inheritance

Operator overloading in a communications channel

Let's look at an example of operator overloading. In communications, sometimes a framing channel is identified by a tuple of identifiers. This is a good candidate for abstracting a simple built-in type. Below is an example of a class `line_id` that illustrates this for a variety of time-division multiplex (TDM) digital signal (DS) levels:

```
class line_id {
  public:
    line_id(uint8, uint8, uint8); //build using specific
      ds3,1,0
    line_id operator++() {//pre-fix
    uint8 max_ds3_(dictionary::find ("max_ds3_", 12);
    uint8 max_ds1_(dictionary::find ("max_ds1", 28);
    uint8 max_ds0_(dictionary::find ("max_ds0", 32);
    if (++ds0_ >= max_ds0_) {
      if (++ds1 >= max_ds1_) ++ds3_;
    }
    assert(ds3_ <= max_ds3_);
    return *this;
  }
    line_id operator--( ) { //pre-fix
    if (!ds0_ ) { //was 0
      if (!ds1_--) {
        assert (ds3_);
        --ds3_;
      }
      else --ds0_;
      return *this;
    }
    bool operator==(const line_id& rhs) {
      return(ds3_ == rhs.ds3_) && (ds1_ == rhs.ds1_)
            && (ds0_ == rhs.ds1_);
    }
  private:
    uint8 ds3_
    uint8 ds1_;
    uint8 ds0_;
};
```

Here it was decided that only `operator++()`, `operator--()`, and `operator==()` were to be provided for the `line_id` class.

Considerations

One last point about operator overloading. C++ supports both global functions and class methods for operator overloading for some functions. The choice of which to use is relatively simple. For operators that act on only one instance, such as `operator++()`, you almost always put the operator inside the class, because this can be much more efficient.

For operators that act on two instances, such as `operator-()`, put the operator inside the class only if the operation is nonsymmetric. For example, most coders assume `operator+()` will give the same result, regardless of the order of the parameters. Consequently, the operator should be in the global scope.

Operator overloading is a very powerful feature of C++. Use it on the small, common classes in a verification system.

Creating Classes—
What the Compiler Will Do for You

When you declare a class (or even a `struct`) in C++, four generic methods are automatically created. They will be used together if you do not define your own versions in your class. These four methods constitute the *canonical form,* and you must consider them when you define a class. Many times canonical forms are sufficient, but there are times when you can't use them.

The four methods created are the *constructor, destructor, copy constructor*, and `operator=()`. Anytime you declare a class, it is as though you typed the following:

```
class scientist {
  public:
    scientist();
    ~scientist();
    scientist(const scientist&); \\copy constructor
    scientist& operator=(const scientist&);
};
```

The canonical constructor is called the *no-argument constructor*. This is needed if, for example, you ever make an array of objects of the class. By default, the constructor calls the no-argument constructors for all data members. If you define any constructor, the compiler will not generate the no-argument constructor.

A destructor will be generated that, by default, calls the destructor of all data members. While this is fine, if you ever derive from the class, you must declare the base class destructor as virtual, so a common technique in most classes is to define your constructor as virtual like this:

```
virtual ~my_class() {}; //assumes nothing to do in dtor
```

The last two methods, called the *copy constructor* and `operator=()` have to do with how an object is copied. The copy constructor is used when an object is initialized with an existing object, as in the following example:

```
scientist newton(); //default constructor
scientist einstein(newton); //copy constructor
```

The most common example of this is the passing of an actual parameter to a function or method. In this case, a new, temporary object is created on the stack and its copy constructor is called. (As an aside, this is why most calls use `const & scientist` as a parameter. This is, in effect, a pointer that cannot be null. Because it's really a pointer, the copy constructor is not called, because no new data are created.)

The other method used to copy an object is `operator=()`. This is called when one object is assigned to another like this:

```
scientist darwin = newton;
```

Note that it is necessary to provide your own copy and `operator=()` methods if your data contains a pointer to allocated data. The default copy and `operator=()` methods will only copy the pointers to the data (because the pointer is the data member). This means that after the default copy constructor is called, the two objects will both think they own the allocated memory and will "double delete" the data in the destructor. In cases like this you need to make sure that the new class (in our case `darwin`) gets a copy of the actual memory, by providing your own methods for the `scientist` class.

An interesting technique is to put the copy constructor and `operator=()` methods in the private scope and not provide an implementation. This prevents an instance of the object from ever being accidently copied. This technique is used when there is only one instance of the class, or when you only pass pointers to any instance of the class.

> *Remember that C++ will create four methods for you each time you declare a class—unless you declare your own versions.*

> *Sometimes when a class lacks a sensible implementation of these default methods, declaring the class in the "private" scope is a good way to enforce intent.*

Advanced Class Techniques

You should be able to create a well-designed verification system by using the techniques presented in this handbook. However, there are some times when it is appropriate to use different and, in some sense, more advanced techniques. Advanced techniques are neither bad nor good; you just need experience to know when to use them. This section introduces some of the advanced techniques and features of C++.

C++ supports having multiple bases for an inherited class. While this is not that uncommon, it should not be your primary focus. Architect your classes with well-defined interfaces, so they will use multiple inheritance if that makes sense.

C++ supports virtual base classes. These are useful if you are using multiple inheritance and need to make sure there is only one copy of a base class. The default C++ behavior is to replicate the class each time it is encountered. In this way, there is no electricity between the various instances. This is fine for most cases, but sometimes a class is used to manage a single resource. By using the `virtual` keyword for a base class, C++ makes sure that only one instance exists in the combined object.

C++ supports exceptions. Throwing an exception, by definition, is the construction of an object followed by stack unwinding. The stack is unwound until either an acceptable catch handler is found, or the top of the stack is reached. Exception handling is generally not needed in verification systems, because simulations don't need to handle unexpected user or algorithmic actions gracefully.

Another advanced technique is to create your own templated classes and methods. Templating is a useful technique to support generic programming, but it defines an interface that is more subtle than an abstract base class. This technique is used sparingly in the handbook.

C++ comes with a standard library of algorithms. Although designed to work with the templated container classes, they will work for other classes and templates that are properly created. The algorithms are rich and varied, although this handbook uses only a few of the standard ones.

C++ has a wide array of techniques. As you gain experience, it becomes more obvious where they can be of benefit.

Summary

• •

This chapter covered a rather broad range of topics. We talked about how to look at the verification environment and see that the nouns are classes and the verbs are methods.

How much "electricity" your design needs was covered next. The basic levels of electricity can be seen as defines, the "if" test, classes, namespaces, operator overloading, pointers, virtual functions, and finally templating. Each step increases the complexity, or electricity, of the code. The idea is to use only the minimal amount of electricity needed.

We then moved on to meta-class-level concepts, such as global services and static methods. Here the idea was that sometimes things really are global.

Class inheritance was looked at in detail. Inheritance is an extremely important OOP concept. By using class inheritance, you can both enforce intent and extend an implementation.

When you create classes, it's important to remember that C++ will write four methods automatically. Be aware that you may want to provide your own implementations.

We ended the chapter with a glance at some of the more advanced capabilities of C++.

For Further Reading

As we stated in the For Further Reading section of the Why C++? chapter, there are no "good" books on how to learn C++. Skim many, buy a few, and copy code where you can.

OOP Connections

Oh what a tangled web we weave,
When first we practice to deceive.
Sir Walter Scott, Marmion. *Canto vi, Stanza 17*

Connecting classes together is more important than the classes themselves. How can this be? It is so because, by definition, the connecting
of classes involves jumping around in the code base. Managing this is
mentally more difficult than simply managing the code within a class.
For example, when you see a pointer to a class in a method call, you have
to think about why the method needs that class. The answer depends upon
whether the system is a tangled web or a well-architected series of
connections.

Often you have to find the header file for the class and go look at the
implementation of the method. In the worst case, the code make no sense
whatsoever, even after you stare at the implementation. In the best case,
the connections are obvious, such as when a test gets a pointer to a
testbench.

Overview

In hardware design we connect modules together and worry about clock domain crossings. With verification, we connect instances of classes and methods together and worry about crossing the threads of execution. In addition, the connections in verification may be temporary (for example, they are used only within a method), or they may be permanent (for example, when a constructor takes in a pointer to a logger and stores it in a data member.

Connections in your code can either form a spider web of complicated and confusing relationships, or they can be a highway, with well-defined points that connect to other roads. Recall that one person's web is another person's highway, so picking the right connection technique may not be universally appreciated. There are many connection techniques and, as a result, trade-offs to be made, as this chapter shows.

We first discuss the various types of connections, then look at implementations of these connection types. We then present the simpler connection types first, increasing the complexity of the connections as the chapter progresses.

We first look at how to classify connections. The type of connection is evaluated according to how much information one class has about another. At one end of the information scale, classes have no mechanism to determine whether any other class instances are connected. At the other end of the scale, a class has a pure virtual method that must be implemented to make the connection.

The idea is to build the appropriate type of connection for the problem at hand. Too loose a connection makes code unnecessarily complicated. This because loose connections make few assumptions about the other side, which in-turn makes tracking events harder. Connections that are too tight, on the other hand, may make code harder to adapt.

While these connection techniques are general, some of them can be used between verification components operating in different threads. Why bring threading into the discussion? In verification systems many events need to happen in parallel. This is normally done through threading. With threading, however, comes a set of problems related to accessing common data. How can a thread be given sole access to a common resource? How can one thread synchronize with another? To solve these problems, *thread-safe connections* where created. This chapter will show techniques to cross thread boundaries.

How Tight a Connection?

Once a verification system has been divided into classes, the next step is to think about is how tight the connections among those classes should be. As we have seen, this is a sliding scale, with trade-offs in complexity. A loose connection creates good flexibility but more complex code. A tight connection is easier to understand but harder to adapt when changes occur. As a result, each side of the connection must make assumptions about the other.

Let's look at two points on this scale. Consider a generic data generator and a data checker. An obvious way to connect these two components is to have one component have a pointer to the other. For example, you could code the connection like this:

```
class data_checker {
public:
  void note_data_generated(uint32 some_data);
};
class data_generator {
public:
  data_generator(data_checker* checker);
};
```

Then you can use the checker and generator like this:

```
data_checker* checker = new data_checker();
data_generator* gen = new data_generator (checker);
```

The pointer example above is considered a *tight* connection. This is because the actual name of one class is given to the other class. Tight connections are obvious and direct, yet they are not always appropriate. They make the code brittle and difficult to modify if the assumptions about either class change. Tight connections are, however, the most commonly used and most appropriate for the common interconnections.

Note that the situation could have been reversed, with a pointer to the `data_generator` given to the `data_checker`. However, in practice the assumptions regarding the number of interconnections are not the same. Often there are several different types of generators, yet usually there is only one checker for a given interface or feature. The number of connected instances is something to think about when you connect classes. It is easier to have many objects point to one then the other way around.

The looser a connection is, the fewer the assumptions are that can be made about it. For example, to continue with the above example, one could instead use an intermediary object to manage the connection. The authors call this a *channel*. In this case, you can give the channel object to both the checker and generator, as follows:

```
class channel {
  public:
    void put_data(uint32 data);
    uint32 get_data();
};
struct data_checker {
  data_checker(channel* expected);
}
struct data_generator {
  data_generator(channel* output);
};
```

Then you can use them like this:

```
channel* a_channel = new channel();
data_checker checker = new data_checker(a_channel);
data_generator gen = new data_generator(a_channel);
```

This is a *loose* connection, because the generator does not know that a checker exists. The generator simply generates data for the channel.

Here is an interesting implementation complexity brought about by our new channel connection. What does the checker do if the `channel::get_data()` has no data? Questions such as this are not necessarily a bad thing; if they are asked in the early coding phase, the resulting code tends to be well-thought out. Channels are an important interconnect technique, and are discussed in detail in a later section.

Sometimes tight connections are appropriate, but at other times looser connections give the appropriate flexibility.

Types of Connections

Now that we have talked about the tightness and looseness of a connection, let's look at the two basic types of connections. One is the peer-to-peer connection, the other the master-slave connection.

Peer-to-peer connections

The peer-to-peer connection occurs when a group of modules are all able to communicate with each other at any time. They may be arranged in various topologies, such as a ring, star, or bus. This type of interconnection usually follows a message-passing scheme and can be tricky to debug.

The Controller Area Network (CAN) protocol is an example of a peer-to-peer interconnect. Any device can initiate a transfer, and the message has the priority, not the sender.

Peer-to-peer connections are not often used in verification systems, because we tend to design systems with a controller in mind. This type of connection is discussed in the next section.

True peer-to-peer connections allow multiple masters and shared communication.

Master-to-slave and push-vs.-pull connections

Contrary to the peer-to-peer connection, most verification components communicate in some sort of unbalanced connection, such as *master-to-slave*. The master initiates an action—either pushing some data to the slave, or demanding/pulling some data from the slave. In either case, the slave must respond.

A *push* connection occurs when one module tells another module to take some data. A common example of this is a generator putting some data into a queue for a BFM to send. The master, in this case the generator, is at a higher layer of abstraction, as opposed to the BFM agent, which "simply" directs the BFM to execute the transaction.

The *pull* form of the master-to-slave connection occurs when one module calls another to get some data. For example, a generator might need to combine data from several sources to form a complete data packet. A specific example of this is when several logical channels share a physical interface. Both the UTOPIA[1] interface and USB interface use this approach.

The appropriate choice of push versus pull is situation dependent. If a connection seems awkward, often reversing the direction of the connection greatly simplifies the code. The general rule is to minimize the number of connections, as well as the assumptions about the connections. If you are uncertain about which type to use, bias your design towards push connections. This is because the decision regarding what to do with the data (for example, whether to send data through a chip interface) is often simpler and of a lower abstraction level than the generator of the data.

Note that at the monitor level, the push is from the monitor towards the checker, because the monitor does not need to know about the recovered data's eventual use. Again, the idea is to minimize the assumptions about an interface.

The following sections are all generalizations of the push/pull interconnection technique.

Most class connections are either push or pull. If the code seems clumsy, try reversing the direction of the connection.

[1.] Universal Test and Operations Physical Interface for ATM.

Two Tight Connection Techniques

The tighter the connection, the simpler the connection tends to be. This is because the techniques for these connections usually name the class or method that is to be used. This is appropriate for a large number of the connections in a verification system. Let's look at two of the most common techniques.

Using pointers

Pointers in a project are as common as ones and zeros. So why talk about them? Pointers can either be just for data, or they can be used to connect different classes. Specifically, in certain cases it's more reasonable to have two classes implement a task than one. This is common when you are crossing abstraction layers. In one case, one side of the connection has some data but does not want to know how the data will be used. In another case, one side of a layer needs more information to complete a task. In both these cases, it might be reasonable to express the "other half" of a task as a separate class. To clarify this, let's look at a specific example.

Suppose a monitor can gather some data, but because we want to separate the data gathered from any processing of the data, we decide to use a pointer to another class. Here, the monitor could take in a pointer to the other class to "handle," or manipulate, the data, as shown below:

```
class handler {
  public:
    virtual void check(int data);
};
class child_monitor {
  public:
    child_monitor(handler* p) : handler_(p); {}
    void data_received(int d){handler_->check(d);}}
  private:
    handler* handler_;
};
```

As we can see, using a pointer to another class minimizes the assumptions regarding what happens to the data.

Consider using pointers when abstraction layers must be crossed. Be careful of what each side of the layer "knows."

Using inheritance

Another way to pass the results of one function to another is to use *inheritance*. While inheritance is a very tight form of connection, it can be very clean and provide good separation between roles and responsibilities. Inheritance can be used as the initial connection, while a looser technique can be used to complete the connection. (This two-step approach to connections is used in almost all the examples in Part IV of this handbook.)

But let's look more closely at the concept of using inheritance for connections.

As an example, consider a base class that contains a verification component's algorithms that either consume or generate data. With this technique there would be pure virtual methods to create or use the data; an inherited class would be responsible for providing or consuming the data.

Let's look at how you might create a generic `checker` base class and how you could use inheritance to make the connection to a "real" checker.

```
class checker {
  public:
    void start {
      vout log("Checker");
      while (true) {
        data expected = next_expected_data();
        data actual = next_actual_data();
        if (expected != actual) log << teal_error
            << "Expected: " << expected
            << " != Actual " << actual << endm;
    }
    virtual data next_expected() = 0; //the connection
  virtual data next_actual() = 0;//the connection
  }
```

Now you can define a "real" checker:

```
class uart_checker : public checker {
  public:
    uart_checker(uart_generator* g, uart_monitor* m) :
            generator_(g), monitor_(m) {};
    virtual data next_expected() {return generator_-
        >next();}
    virtual data next_actual(monitor-
        _>wait_for_next_data());
  private:
    uart_generator* generator_;
    uart_monitor* monitor_;
}
```

Note that the base class, checker, has no knowledge of how the data are gathered. This type of connection can be good for separating the responsibility of getting the data from the responsibility of checking the data.

Inheritance is well-suited for the initial connection to the classes outside the base class. In this example, the uart_checker is connected to a generator and a monitor and is waiting for the data. A different implementation class could connect to queues of data. Still another implementation class could use the generator, but for the monitor it would wait for an event and then read the received data from the HDL.

The same two-step technique could be used to minimize the connection assumptions that need to be made for a BFM or monitor. A pure virtual method could be used to consume the data. One subclass might put the data into a channel, while another might filter for special packet processing and then send the data to a specific checker on the basis of this processing. The XON/XOFF processing of the UART interface is a good candidate for this type of connection, as is the processing of Ethernet multicast packets.

> *Inheritance may be a way to defer the specific*
> *interconnection mechanism and thus be useful— or it can add*
> *complexity, if only one type of interconnection subclass is*
> *used in practice.*

Threads and Connections

As we discussed in the overview, verification systems use threads in proportion to the concurrent activities in the chip. Therefore, it's natural to build verification systems that mirror this parallelism. To make the connection between the independent threads, we need a connection that can pass data between threads. This is called a *thread-safe connection*. We'll talk about the base thread safety mechanism, the event, and then move on to fancier thread-safe connection techniques.

Events—explicit blocking interconnects

We now shift our focus from the general types of interconnect to those that can cross a thread boundary. This is important because we use threads often in verification.

Most threads are synchronized by an underlying "wait and signal" mechanism. This is done by an object called an *event*. An event blocks a calling thread until another thread signals that the event has occurred. This technique is the building block of most higher-level interconnect mechanisms, but it can be useful as a technique by itself.

One thread waits for an event to be signaled, while another thread signals the event. This is a good mechanism for coordination, because the waiting thread needs to know only the name of that event. Note that the signaling of the event generally indicates that the other thread has entered a desired state being waited on.

For example, consider a protocol error generator. This error generator forces the wires of an interface to an illegal value during a specific phase of the wire protocol. In order to achieve this, the error generator needs to know the phase of the protocol. It can do this by explicitly monitoring the wires, but it is better to separate roles and responsibilities by using a separate monitor. The monitor is responsible for providing events that trigger on the beginning of the different parts of the protocol. The generator just has to decide what part of the protocol to corrupt, and when, then wait for the specific monitor event. After the event is signaled, it can force the wires into an illegal state until the next protocol state is signaled.

This is shown in the example below, which creates errors in the Cyclic Redundancy Check (CRC) phase of the protocol.

```
class protocol_monitor {
  public:
    event crc_phase_begin; //set when crc is detected
};
class crc_corruptor {
  public:
    crc_corruptor(protocol_monitor* pm) :
        protocol_monitor_(pm) {}
    void start() {
      protocol_monitor_->crc_phase_begin->wait();
      //Now force the wires to corrupt the crc
    }
  private:
    protocol_monitor* protocol_monitor_;
}
```

After the `crc_corruptor` is hooked up to the `protocol_monitor`, the `crc_corruptor` is started. The latter then waits for the signal from the monitor and then trashes the CRC. We have separated the protocol specifics from the desired action.

Note that there are no data other than the fact that the event occurred. The fact that the event occurred is all that is needed in this example. However, this can be limiting, as threads often exchange data on the basis of some coordinating event. This issue is solved in the following sections.

Events are useful as a form of a connection, because the data exchange is minimal. Make sure that the triggering of the event is all that is really needed.

Hiding the thread block in a method

Instead of a just using an explicit event for the connection, consider hiding the event behind a class method. This is called a *blocking method*, because the method blocks the calling thread. By using this technique you can associate data with the event. Also, the event is now abstracted into a method.

Moving to a blocking call makes coding sense, because the choice of using an event, an HDL wire, or a set of events is now up to the implementor of the class. It also simplifies the interface, because method calls are a standard way of communicating. The fact that the call is blocking can almost be an implementation detail. This can make the code clearer or more confusing, depending on whether a user of the class can reasonably expect that the code will block. Sometimes this blocking can be implied by the method name.

As an example, assume that a protocol monitor has the method `wait_for_start_of_frame()`. Because of the "`wait_for`" in the name, one can assume it will block the calling thread. Now the monitor is free to implement the method in any way that best fits a specific design. Perhaps it has an internal event called `start_of_frame` that is triggered by an internal thread. Alternatively, it could have an internal boolean variable `start_of_frame_`, and poll it on the positive edge of a clock. Another implementation might be an internal state machine and a single event that indicates a change in the state. The point is to separate the interface from the implementation, minimizing the implementation assumptions.

Another variant on the blocking method is to use an overloaded method, commonly called `wait()` or `trap()`. There will be several `wait()` methods, each with a different pointer to an object that specifies the event desired and the data to be returned.

Continuing with our monitor example, suppose the monitor supported three blocking methods: waiting for start of frame, waiting for start of data, and waiting for the data packet to complete. The following is an example interface:

```
class start_of_frame {public: uint32 frame_number;};
class start_of_data {}; //a class, the only data needed

class data_complete {
  public:
    std::vector<uint32> data;
    uint16 crc_16;
};

class monitor {
  public:
    //C++ uses parameters to pick the right wait method.
    void wait(start_of_frame*);
    void wait(start_of_data*);
    void wait(data_complete*);
};
```

This method has the advantage of using a convention to show that the methods are related and that blocking semantics are used. Note that this method is not clear if inheritance is expected. This is because an inherited class that provides a wait() removes all the base class wait() methods—unless they are explicitly brought back into scope.

> *Using a blocking method is often better than using an explicit event. Make sure the method name conveys the block, if the block is not just an implementation detail.*

Fancier Connections

The connection techniques discussed above provide a good basis for fancier, more complicated connections. So why did coders invent these fancier connections? The techniques discussed below are combinations of the basic ones, and are used to express the coder's intent better. Although you might not use all these techniques all the time, it is good to have them in your bag of tricks.

Listener or callback connections

Sometimes you do not need a two-way connection, but just need to "listen in" on another object. A technique for this type of connection is called the *listener,* sometimes also called *callbacks.*

Why two names for the same thing? Programmers often used the term "listeners" when they are viewing the architecture from outside of the class to be listened to. Programmers use the term "callbacks" when they are speaking from the other side, from the class with the interesting data. Confused? Don't worry, this is not a technique that we recommend or use often, as we explain below. We'll use the term "listeners" throughout this section.

Listeners are objects that are called at specific points in another object's methods. Often the two objects are unrelated, although often a pointer to the calling object (the one with the interesting data) is passed.

Remember our monitor example, which had a "start_of_frame," "start_of_data," and "data_completed" thread synchronization points? Instead of an object representing just the interesting synchronization points, there could be generic listener objects that would be called at many points in the monitor's state machine. An interface could be like the following:

```
class monitor; //forward declare allows action to use a
    pointer
class action {
  public:
    virtual do_action(const monitor&) = 0;
      //perform action, with monitor state as needed.
}
class monitor {
  public:
    const uint& current_frame;
    void add_start_of_frame_listener(action*);
    const data* current_data;
    void add_start_of_data_listener(action*);
    void add_data_complete_listener(action*);
}
```

Note that this is functionally equivalent at a high level to what we had in the previous section. However, in this case, instead of a blocking method, an object's `do_action()` is called. This may make the intended task easier or harder, depending on the task.

If the listener's task is relatively self-contained, as with incrementing a counter, this technique is straightforward. If, instead, the task is to implement some high-level algorithm and that code `case`'d on several of these state changes, the multiple listeners needed would be an extremely clumsy way to express the algorithm.

This technique can also be used when the author of the original code cannot allow an inheritance-based interconnect, or you cannot get access to the source code.

There are several variants of this technique. The listeners could each have their own class, be separated into pre- and postmethod listeners, or act as filters for some data.

> *As a rule, use the listener/callback interconnect only when you are relatively sure where to put the callbacks. In addition, ensure that there are many simple, loosely connected actions.*

Channel connections

A *channel* is a connection technique that manages a queue of data between two or more objects. It is a fancy way to pass the data between classes. The technique is a loose form of connection, because both sides interact through an intermediary.

Channels usually handle the crossing of thread boundaries. One verification component places data into the channel, while another component—possibly at a later simulation time and in a different thread—consumes the data. One side of the connection has no knowledge or assumptions about the other.

While channels can complicate debugging, there are many situations where they are a necessary trade-off. For example, a generator usually has to send the data to both a BFM and a checker. This can be accomplished by having a channel replicate its data into two channels. As far as the generator knows, it is sending data into only one channel.

Another example occurs when two or more verification components want to send data to a third component, such as a transactor. You can create a channel that takes input from any number of channels and merges the data into a single channel.

Note that a common channel behavior is that the thread consuming the data waits for the data to become available. In this case the channel implementation uses an event.

A channel is specific to the data it contains. In C++ terminology, a channel is a *container class*. Container classes are good candidates for templating.

Channel connections are very useful in verification. They are a loose, thread-safe mechanism for connecting a number of components together.

Action object connections

Sometimes just having a channel with data is not sufficient for what you want to express. In this case you need to have an object that can "do" something. Maybe the object makes some configuration calls to a BFM, or it just sends a burst of data. In any event, this more active connection is called an *action object connection*.

This method of connection combines the channel and the listener techniques. Even though the data in the channel connection are objects, the idea here is that the channel generally contains passive data. With active object connections the channel can contain both control and data. This can lead to code that is obscured and hard to find, and therefore hard to reason about.

With the use of action object connections, there is a channel or queue of objects, each with a listener type of action object, and a single method: `void do_action()`. Various objects create these action objects and place them in a command queue. The owner of the queue pops the action object off and calls its `do_action()` method. This method usually calls some configuration methods in the target object and probably also a `put()` or `get()` method. In this way arbitrary sequences of control and data can be queued.

Action object connections can be used to synchronize control and data—a good thing when you want to encode configuration settings, and have a generator create sequences of configurations and data using those configurations. However, although a large number of chips can accept configuration changes with data flowing, make sure that capability is intended to be used in production software.

The authors have rarely used the action-object connection technique, but it has proved useful for complex sequencing problems.

Using action object connections in a channel for the complex sequencing of a chip may be appropriate for testing a CPU or graphics chip, but it is probably overly complex for testing an interface or feature.

Summary

In this chapter we have explored various ways to connect classes. Which technique should you use? It all depends on the problem you are trying to solve. Different connection techniques have different trade-offs. In general, try to use the tightest connection you can, because that will be the most obvious connection.

We talked about the two most common forms of connection, pointers and inheritance. Note that the pointer is an instance-based concept, while inheritance is a class-based one. The significance of this is that the class-based technique is static, and thus slightly simpler.

Using events is a good technique to let one thread know when another thread has changed state. Events are fundamental to thread-safe connections.

You may decide to hide the event by using a blocking method call. This is a good technique for loose connections, such as between monitors and error injectors.

Not surprisingly, the most common connection technique (besides pointers) is to pass queues of data. A channel, a common implementation of this technique, crosses thread boundaries and separates the producer of some data from the consumer. It also allows for clever techniques, such as replicating the data to another channel and filtering the data to add errors.

The final technique we looked at was called active object connections. These are used when you need to mix control and configuration with the data. As with listeners or callbacks, this approach can be a slippery slope. Although everything can be expressed in a mixed control and data channel, just make sure you only use what is needed.

For Further Reading

- On the topic of connections between levels of abstraction and within a level of abstraction, *Software Engineering: A Practitioner's Approach,* by Roger S. Pressman, has a several relevant sections. The fancy term for the connections is called "cohesion and coupling."

- The authors are aware of several books and papers on different connection techniques. None are landmark or stand out as "the best" way. As with learning C++, this is more of an impedance matching issue, with some books and papers better matched to your learning style and experience level than others.

Coding OOP

Beauty is in the eye of the beholder.
Common paraphrasing of Plato

Coding is a personal endeavor. For many of us it's similar to creating art, and as with any art, there are many styles—some loved, others detested. Why is this relevant to coding? Well, because unlike the case with art, our code cannot stand alone. We are in the interesting position of creating art that, by definition, must work in a community.

No engineer intends to create complicated, stand-alone code. This chapter shows techniques, tricks, and idioms that you can use to communicate your intent. When your code is clear and transparent, other engineers can more easily understand, and appreciate, your intent. Code that is appreciated is more likely to be used appropriately, adapted, and, most important, integrated well with the rest of the system.

Overview

This chapter shows some of the coding techniques we can use to create our art. This, the last of the OOP chapters, talks about the coding going on inside a class. Of course, what's going on in a class is related to the class structures and interconnects around the code, so we will not limit our discussions to the lines of code in a method. Rather, we focus in this chapter on coding.

Our first focus is on "if" tests, with a discussion on why this necessary coding construct complicates the code. We'll show some ways to minimize these "if" tests.

We then discuss ways to get your point across, using coding tricks and idioms. We also look at the touchy subject of coding conventions, and try to point out where they help and where they hinder.

Finally, we look at templating and how it can be useful. The authors realize that we have covered templating in a few of the previous chapters, but here we take one last look at the C++ template library, as well as one more example of when to write your own template.

The previous part of this handbook used many different techniques, so we figured that one more look, concentrating on when to use individual techniques, might be useful.

"If" Tests—A Necessary Evil

The fewer "if" tests a segment of code has, the easier the code is to understand. Code that just does step one, followed by step two, and so on is inherently easier to reason about. An "if" test, by contrast, causes us to think some more. Is the condition true? Will the code set something up that I have to remember? Where was this condition set?

This section looks at ways to minimize "if" tests in your code. Of course, there will always be "if" tests in code; the goal is to find the place to put them so that their presence is reasonable and maybe even expected.

To put this more generally, writing code consists of procedural statements and changes in control. The mix of these components two dramatically affects the complexity of the code and its adaptability. Procedural statements are just unconditional expressions followed by a "; "—for example, in function calls or mathematical statements. Because the processor always executes the statements in order, it is not difficult to understand the control flow.

By contrast, changes in control increase the "mental state" that must be remembered. One must now consider two paths through a block of code. These changes in control can be looping constructs, "if" tests, "switch" statements, "?" operators, or goto's. This section is primarily concerned about "if" tests. Loop constructs can be considered as combinations of an "if" test and a goto. "Switch" statements can be considered as restricted implementations of "if" tests, so they are simpler, but still not as simple as procedural code.

The question mark operator is somewhat like an "if" statement, but, in practice, is best relegated to assignment statements, such as in a = (b > 4) ? 10 : 62. In this case, because it is a common idiom, the operator does not increase the mental complexity of the code.

"If" tests, while necessary, can complicate the code. Using them where they make sense is tricky.

"If" tests and abstraction levels

Almost all algorithms have "if" tests as part of their definition. Algorithms at the BFM/monitor level are probably best implemented with these "if" tests. However, as the abstraction level increases, the number of "if" tests should decrease. This is because "if" tests make it harder to reason about the code. At the higher levels, the code should be fairly straightforward.

If there are "if" tests at the higher levels, they should be more of a "presence or absence" test, as in the following:

```
if (scenario::provide_jtag_traffic()) {
    jtag_stimulus = new jtag();
}
```

These kinds of "if" tests still raise questions, but questions that should be more easily answered.

At the high level, the "if" test should be rare.

"If" tests and code structure

Be careful about using "if" tests whose body exceeds a few lines. The body is the part of the code that is affected by the "if" test. This can complicate the understanding of the code, because the structure of the algorithm is obscured. The authors have pondered many examples of this.

Sometimes the "if" tests are preventing the last inner code block from executing. A common example of this occurs when you have a series of conditions that must be met (perhaps stages in some protocol) before you execute the inner code. In this case, consider using return statements on the inverse of the conditions to weed out these large blocks of "if" tests.[1]

Let's look at a specific example. Here is a function that has to process a packet of data:

```
void my_class::handle_packet(const packet& a_packet)
{
if (a_packet.valid()) {
  if (i_am_enabled_) {
    if (we_are_in_sync_) {
      packet_count++;
      if (system::error_level(1)) {
        //...the algorithm here...
      }
    }
  }
}
```

[1.] This is yet another case where the academics differ from the industrial coders. You may have been taught "there is only one return from a function," but this rule—like all rules—works best if applied with intelligence, not dogma.

```
      }
    }
```

Instead, consider this reorganization of the code with a negation of the original condition:

```
void my_class::handle_packet(packet a_packet)
{
  if (!a_packet.valid()) return;
  if (!i_am_enabled_) return;
  packet_count_++;
  if (!system::error_level(1)) return;
  //...the algorithm here...
  }
}
```

In this case the algorithm is clearer. There are some preliminary tests, and, if they pass, the main algorithm is performed. This allows you to forget the "bad" cases and concentrate on the core algorithm.

However, this technique of testing is not always appropriate. As with most things in coding, it's a trade-off. The counter-argument against testing for "not true" is three-fold: (1) testing for the negative is sometimes counterintuitive; (2) multiple returns complicate an algorithm; and (3) there is the fact that, because you do not have indentation in the code syntax to help you, you must remember what tests have passed as you read down the function.

> *Avoid long "if" tests. Also, sometimes "if" test reversals make the code easier to understand.*

Repeated "if" expressions

There is a class of "if" tests that are particularly damaging to reasoning about the code. These are the tests whose expression has already been tested. These tests make us keep rethinking a condition and make us nervous as to why another check is needed. Can the condition have changed from the first test?

In addition, there is a danger that the person reading the code does not see all the repeated expressions and misses some when the expression must be modified.

There are two solutions to this situation, assuming the repeated expression is in a single class or function. One solution is to encode the results of the test into a boolean data member and then use this data member instead of the class. The other method is to rethink the algorithm. It is quite possible that the algorithm needs to be "warmed over," recrystallized, or converted from push to pull.

Be extremely cautious about repeated "if" expressions in different objects. This is almost always a design mistake, because the algorithm is being spread out across several classes, making it difficult to change it.

The authors once worked on a piece of code designed to power down the units of a chip that were not being used. However, once we changed the code, the test hung at the end. It turned out that, by powering down the logic block, the chip would not respond to a register read and would hang. In retrospect, this behavior made sense.

The code that was doing the read was confirming that the module had not raised any errors. This was fine, but the test for "module in use" boolean was repeated in another part of the system that the authors had missed.

In the end, this bad style of widely distributed, repeated "if" tests was not all bad for this project. The incorrect "if" test showed that, if software accessed a register of a powered-down module, the chip would hang. The designers ended up adding a simple watchdog timer on the internal bus.

Retesting the same condition test is clumsy, complicated, and error-prone.

"If" tests and factory functions

So "if" tests are dangerous. How do we make as few of them as possible? One way of minimizing "if" tests is to encode the different algorithms in a class hierarchy with a common base class, and use a single function to build these inherited classes. This function, called a *factory function,* is given information to help it decide what inherited class to create. The factory function returns a base class pointer. The rest of the system would not know the specifics of the actual class; for all the code knows, the object is a base class.

This is a very powerful technique for code adaptability. If the factory function is separated from the class hierarchy, it can be adapted to different situations without the need to change the rest of the system, because the rest of the system knows only about the base class.

This reduces the need for "if" tests, because the mechanism of virtual functions is taking their place. Very OOP!

A factory function example

As an example, consider a task to test a Controller Area Network (CAN) protocol implementation. The testbench consisted of a reference model (verification IP), a hardware implementation (which had three ways to drive CAN, DMA, FIFO, and register), and a hardware-assisted peripheral interface controller (PIC) implementation (another small microprocessor on the same chip). Because the CAN protocol is a multinode protocol, several nodes were created. A factory function was used to return a generic can_node class, even though within the factory function one of five possible inherited classes were built.

The traffic generator did not have any knowledge about the specific CAN implementation connected to it. The generator would simply send random traffic. The checker also did not know about the implementation of the nodes. It would just make sure all the nodes were in the same state (after each bit time) and report any differences.

Here is the can_node base class that the generator and checker had access to:

```
class can_node {//The generic base class
public:
  can_node(const std::string& name, configuration c,
      truss::port<configuration::signals>::pins wires):
      configuration_(c), port_(wires), name_(name) {}

  void init() {init_();}
  void start() {start_()}
  void stop() {stop_();}

  void send_message(const message_data& m)
      {send_message_ (m);}
  void message_received(const message_data& m)
```

```
            {message_received_(m);}
    const std::string& name() const {return name_;}

  protected:
    virtual void init_() = 0;
    virtual void start_() =0;
    virtual void stop_() =0;
    virtual void send_message_(const message_data&); = 0
    virtual void message_received_
            (const message_data&) =0;

    const configuration configuration_;
    truss::port<configuration::signals>::pins port_;
  private:
    const std::string name_;
};
```

The base class is the foundation for all the inherited classes. In this example, it will specify how to start and stop the node, as well as how to get and send messages. The `init_()`, `start_()`, `stop_()`, `send_message_()`, and `receive_message_()` methods are pure virtual methods, which means that the inherited classes must provide their implementation. This makes sense, because in our example we have several implementations. The base class would not know how to interact with the chip.

The `name()` method is a nonvirtual one, with a `const` prototype. This means that the class cannot change any data members when the call is executed. This is a useful technique to separate the accessor and status methods from the ones that change the data members of the class. (The exception to this rule is data members with the keyword "`mutable`, which is generally used for cached data that are logically part of the status of the class.)

The `private` section of the base class has only the name of the instance, and there is no inherent reason that this should be hidden from the inherited classes. Because the name is a constant, the inherited classes cannot change it after construction. Note that one may want to append `_fifo`, `_dma`, `_pic`, or `_vip` to the name, but that should be done in the constructor of the inherited classes.

Now that the base class is defined, the inherited classes, such as can_fifo, can be defined. They are prototyped here, but the implementation is too specific to be explained in this handbook.

A factory function to build the CAN nodes was used. First, here are the inherited classes, without their implementations:

```
class can_node_dma : public can_node {};
class can_node_fifo:public can_node {};
class can_node_register: public can_node {};
class can_node_pic : public can_node {};
class can_node_vip : public can_node {};
```

Now, an enum is defined that can be used to select which specific node type to build:

```
typedef enum {dma, fifo, register, pic, vip} can_type;
```

At this point, assuming that we randomize on what type of node we want to build, a single function can be called to create the specified type of node. This is shown below:

```
//The Factory Function -
// Returns 1 out of 5 possible implementations

can_node* build_can_node(can_type the_can_type,
    const std::string& name, configuration c,
    truss::port<configuration::signals>::pins p)
{
  switch(the_can_type) {
    case dma: return new can_node_dma (name, c, p);
    case fifo: return new can_node_fifo(name, c, p);
    case register: return new can_node_register(name,
                                              c, p);
    case pic: return new can_node_pic(name, c, p);
    case vip: return new can_node_vip(name, c, p);
    default: truss_assert(0);
  }
}
```

Given the above factory function, the top layer of the CAN test just builds the nodes and connects them to generators. Because the chip and the PIC were being developed at the same time, we also had separate tests that built only one of those types and a reference node. It was only the top layer of the CAN test that had any knowledge of the specific types of nodes that were being connected. This was later randomized to test various combinations of nodes.

> *Using factory functions to build a specific inherited class is an OOP technique to reduce "if" tests.*

Coding Tricks

When we code we tend to use patterns that have helped us in the past. This section presents a few such patterns. This section, as does the following section on idioms, shows conventions that have helped coders focus their thoughts and tighten their code.

Coding only what you need to know

This is perhaps the cardinal rule for creating good code. It's based, of course, on the same assumption that creates our profession of verification:

> *Code that is not verified will contain bugs.*

Sure, by using this technique we write code that does not have all the features that every situation needs—but a smaller system is easier to reason about and thus adapt. Remember, if a coder needs to add code to your class, it's because they need it.

Another reason for coding only what is needed is because then the code that exists is at least verified to some level. It's frustrating to work on a method, only to find out after hours of debugging that it was never used and does not work.

> *If a feature is present in some code, it had better be working code. Be cautious about implementing features you will not use today.*

Reservable resources

The majority of hardware has a bus to access configuration registers. As the verification system we write consists of many threads, there is a danger that two threads can start using the bus at the same time. The hardware bus is considered a *reservable resource*, because code must first request access to the resource. The trick is to make the reservation as simple as possible for the code that uses the resource.

The simplest solution is to hide the reservation inside the implementation of the class. This is appropriate in most cases. A simple mutual exclusion (mutex) algorithm can be used for this purpose. A mutex only allows one thread at a time into a section of code. The latter parts of this handbook show examples of using a mutex in a register-access BFM.

Sometimes the hardware can process several requests at the same time. This is probably an implementation detail when used in a full chip test, but it is probably something you need to expose at the unit level test. In this case there are two classes. One class exposes a *key*, such as an integer tag or an instance in the interface. The other class uses this lower-level class and hides the key from the rest of the system.

The concept of reservable resources can also exist solely in the verification system itself. You might, for example, have a DMA descriptor queue and need to allocate and release descriptors. Of course, the hardware actually implements the queue, but the management of the queue is a verification concept.

> *Reservable resources may be an implementation detail, and thus use a mutex internally—or they may be an external property, in which case a "key" must be used. This key can be anything from an integer to an object, depending on how safe the management must be.*

Using register fields instead of hard-coded integers

A large part of what a verification system sets up is the chip registers. It is often a good idea to assign registers by means of their fields. This makes the code clearer and allow fields to be moved among registers with little pain.

There are several ways to do this, ranging from simple `#define`s to creating classes for each register and methods or strings for each field. Each verification team will have its own preferences.

One item to keep in mind when settling on a register and field identification technique is whether transparency of the code to production software or diagnostics is a requirement. If so, the software team must be involved in the decision. Note that, while software is probably just doing something like `*((int*)(0x062896))` to read or write the register, verification probably needs each register read and write at least to be mapped into a function or a `define`, such as `reg_read(address)`. Note that, because a system may have back- and front-door reads, the verification effort may map this to one of several mechanisms. Also, there may be multiple front-door implementations, such as through a processor bus, a PCI, or even JTAG[2] interface. The last part of this handbook shows examples of register access and how they can be mapped in a flexible, but obvious, way.

Registers are an important concept, but don't let them obscure the BFMs or the test.

Using data members carefully

When you start building a class, there is a tendency to make many data members. It is common to see a number of calls that have no parameters, but that use the data members in the class as a shorthand. This is fine when those methods are called from outside of the class. However, for a protected or private method that is called by a public method, consider using the standard C parameter passing instead of a data member.

Here is an example:

```
class a_class {
  int value_;
  int weak_data_member_;
public:
  //called from outside the class, use data member
  void method1(int value) {value_ = value;}
```

---

[2] Joint Test Action Group, IEEE 1149.1

```
    //called from outside the class, use data member
    void method2() {value++;}
    void method3() {
      //Is this confusing?
      weak_data_member_ = value_ + 3;
      method4();
      method5();
    }
  private:
    void method4() {weak_data_member_ += 10;}
    void method5() {value_ = weak_data_member_;}
  }
```

The reasoning is that a data member is a bit like global state and comes into a method whether or not you want it to. As such, it makes the class slightly more complicated. This is fine where it is necessary, but inappropriate if the data could have been simply passed in as a parameter.

The fancy term for all this is *spatial locality*. In our case this means that the data are needed by multiple calls *from outside the class*.

A related fancy term is *temporal locality*, which refers to code that is in different functions, but is *called sequentially* (think `{object1->do_method(); object2->do_another_method();}`).

In general, with spacial locality you want to use data members, with temporal locality you want to use parameters to the calls.

Here is the example reworked to pass parameters (this example has temporal locality, but not spatial locality):

```
  class a_class {
      int value_;
  public:
    //other methods as before...
    //less confusing?
    void method3() {method5(method4(value_ + 3));}
  private:
      int method4(int temp) {return(temp + 10);}
      void method5(int temp) {value_ = temp;}
  }
```

Use data members sparingly. Make sure a data member is needed because of spacial locality.

Coding Idioms

An *idiom* is a fancy word for a coding trick that can be expressed not only as a concept, but also in a well-known code structure. This section introduces some idioms that the authors have found to be useful for building verification systems.

The singleton idiom

Sometimes a class is meant to be instantiated only once, and it has no clear owners. The fancy term for this is *global service*, as was discussed a bit in the chapter on OOP classes. Let's look in detail at a common implementation of this one-off instantiation, the *singleton*. A singleton uses a single static method, called `get()`, to return a pointer (or reference) to this single instance.

Consider the following example:

```
class channel_counter {
public:
  static channel_counter& get() {
    assert(channel_counter_);
    return *channel_counter_;}
  }
  static void start() {
    assert (!channel_counter_);
    channel_counter_ = new channel_counter();
  }
  static stop() {
    channel_counter t = channel_counter_;
    channel_counter_ = 0;
    if (t) delete t;
  }
private:
  static channel_counter* channel_counter_;
};
```

Another common convention for singletons is just to have a global function that returns a reference to the global object. This global function may be put into a namespace if it makes the idiom clearer.

Note that the creation of the internal implementation pointer is a different matter. There are different ways to do this, from automatically creating one on first use, to having a factory function, to having an static `start()` method. Which mechanism to use is a personal choice.

Singletons are a good way to implement a global service.

Public nonvirtual methods:
Virtual protected methods

When you are coding a class, there are often *virtual functions*. These methods provide the implementation of either the whole interface of the class, or perhaps just a few specific details. Your first instinct is to make these virtual methods public, and this might be good. However, sometimes you need to do some basics things first, or perhaps afterwards. How do you guarantee that the pre- or postcode is called?

The trick is to have a *public nonvirtual method* that just does the pre- or postcode and then calls the same named method (with an identifier, such as a trailing underscore) as a *virtual protected method*. This allows any standard preamble or postamble code to be guaranteed to be executed. Sometimes you might want to use this trick even if there is no special code. It's a useful technique to separate an interface method (those with an underscore, or "_") from an implementation method.

Here is a short code snippet:

```
class thread {
public:
  void start() {start_(); thread_count_++;}
protected:
  virtual void start_() = 0;
private:
  uint32 thread_count_;
};
```

In this case the public interface is through the `start()` method. The actual implementation is done through inherited classes by means of the `start_()` method. This allows a reader of a class to concentrate on the public, "nonunderscored" methods. It also allows coders that need to inherit from this class to concentrate on the protected, "underscored" methods. Note that in all the implementations the authors have used, the nonvirtual method is optimized away by the compiler, so there is no run-time penalty for this increased clarity.

With this technique, the nonvirtual public method is firmly in control and calls the virtual method only after performing any desired pre or post actions. Sometimes, though, the very nature of the call expects pre- and postconditions. In this case it is clumsy for the inherited class to have to remember to call the base class method. If the designer of the base class wants to encourage, or anticipates, such usage, it's better to add virtual pre and post methods explicitly.

Here's a code snippet that can be used in this case:

```
class generator {
public:
  void generate_one() {
    __generate_one();
    //code here to do the standard generate_one()
    packet_count++;
    generate_one__();
  }
protected:
  virtual void __generate_one() = 0;
  virtual void generate_one__() = 0;
private:
  uint32 packet_count;
};
```

In this case the "main" method—`generate_one()`—is not virtual, but the pre and post methods are. One convention the authors have used is to write `pre_` and `post_` as prefixes to identify the set of methods. However, the convention that the authors prefer is to name the pre method the same as the main method, but with a double underscore ("__") prepended. The post method is similar, but with a double underscore appended. In this convention, the reason the letters "pre_" and "post_" are not used is that they can interfere with the semantics of the name of the original method (which might be something like `post_process`, or `post_completions`, or `prefetch_data`). As is a common theme in this handbook, the choice is yours.

> *To enforce that special pre or post code is called, use combinations of public nonvirtual methods and protected virtual methods.*

Pointers for change, references for efficiency

The C++ language allows you to use references instead of pointers. The original purpose of references was to allow the compiler to optimize runtime costs for passing a parameter. However, references can also be used for changing parameters without the caller knowing about the change. This is often bad, because it's confusing. For this reason the authors use only constant references, as shown below:

```
class big_data;
class a_class {
public:
  void do_something(const big_data&);
}
```

To the caller it looks as if the data passed in will be copied. However, if the amount of data is large, or copying is not allowed, the call can mark the data parameter as a constant reference. This means that data will not be modified during the call.

By contrast, sometimes you do want to modify a parameter. In this case consider using a pointer, which will alert the caller to the fact that the parameter will change.

The authors realize this may seem like a small issue, but it slows down learning a new algorithm if you have to keep going back to every prototype to see if it may have changed.

> *Consider using* `const&` *data for efficiency reasons, and pointers when you will change the data. Avoid non-*`const` *references.*

Avoiding accessor methods— using constant references instead

If you took a course in programming in the past several years, your instructor might have said never to make data members public. Instead, you were told to create *accessor methods*. Accessor methods are the simple one-line getters or setters that are just in front of the data members.

That advice was just silly. The presence of accessor methods does not change the fact that the abstraction is weaker. In other words, the interface has an assumption that those accessor functions are going to execute in constant time. Algorithms may not work if what was a simple `get()` of a variable now takes a significant amount of time.

Instead of using accessor functions, consider using a `const` reference to an internal data member. This ensures constant time access and makes clear the design intent.

Consider, for example, the following:

```
class request_handler {
  public:
    const int& outstanding_requests;
    request_handler() :
      outstanding_requests(outstanding_requests_) {}
  private:
    int outstanding_requests_; //updates, based on
      internal state
}
```

In this code the `outstanding_requests` can be accessed by anyone at any time. Note that this technique now makes it clear that the abstraction for `outstanding_requests` is weak. However, this weakness may have been the best solution to the problems the system was addressing.

The C++ feature of a `const` reference is a good way to bring out the constant-time access parameters of a class.

Enumeration for Data, Integer for Interface

Enumerations (enums) were introduced in programming languages to make the code clearer. They are more powerful than defines. In fact, in C++ they are rather close to classes.

Using enums when setting up parameters can increase the communication level, but there are a few dangers. One occurs when the enum is "`case`"d, or "if" tested. This is can lead to unexpected behavior when enumerations are added. *Enums should generally be used as a shorthand for integral values.*

To this end, the method that uses the enum should sometimes take in an integer as the formal parameter. Why? Because this allows for future expansion (enums cannot be subclassed) as the integral value of the enum becomes the important part of the method's implementation.

For example, consider a baud rate enumeration:

```
class configuration {
  public:
    typedef enum {b_9600 = 9600, b_19200 = 19200,
                  b_921600 = 921600} baud_rate;
    non_virtual void new_baud_rate(uint32 new_value);
  private:
    uint32 baud_rate_;
}
```

Again, this is one of those things that you were probably not taught in class. You would have been told to define an enum and use it in all parameter declarations. That technique does work a fair amount of the time, particularly if the range of the enums is fixed for all time. However, in the messy world of coding for a living, sometimes we need to be a little more flexible.

> *Sometimes you should define an* enum *in a class, but take in an* int *for the methods. Note that mixing enumerations and integers is not always desirable, as it weakens the abstraction. The idea is to use this technique only when future derivations need it.*

Using the constructor and destructor to simplify the code

C++ automatically calls the destructor when an object goes out of scope or is deleted. This can be used to simplify many algorithms. In general, you can pack the tedious work into a little object and let the constructor and destructor handle the bookkeeping.

One example of this is *sentry objects*, which do something in the constructor and undo it in the destructor. For example, a mutex sentry class can be used to get and release a mutex automatically:

```
class sentry {
  public:
    sentry(teal::mutex* m)  : mutex_(m)
      {assert(m); m->enter();}
    ~sentry() {mutex_->exit()}
  private:
    teal::mutex* mutex_;
}
```

It can then be used nonintrusively in a class, such as a BFM, to make sure only one verification thread at a time accesses the bus:

```
class serial_bus_access {
  public:
    serial_access() : mutex_(new
      teal::mutex("serial_access example")) {}

    void read (uint32 adr, uint32* d){
      sentry sentry(mutex_);
      //code to use a hardware bus one thread at a time
    }
    void write (uint32 adr, uint32 d){
      sentry sentry(mutex_);
      //perform write under mutex protection
    }
  private:
    teal::mutex* mutex_;
}
```

Note that you can combine this technique with declaring objects only when you need them. In this way you only acquire the mutex just before you need it.

Use the automatic construction and destruction of an object to your advantage. Avoid creating all your objects at the top of a function. They will all be constructed and destructed automatically, even if you return from a method and never access that object.

Using a boolean to indicate success

Here is another technique for your bag of tricks. In some algorithms there is a series of allocations that are performed between several "if" tests. The general idea is that an algorithm allocates some storage, and then does some work and allocates some different storage or objects. Assume that at some point the algorithm encounters an error and needs to return. Remembering what deallocations to perform, depending on where the algorithm found an error, is tedious and error prone.

For example, a transfer of a USB control endpoint requires several allocations and deallocations of chip memory to complete. An error injector may randomly inject errors during this sequence, causing the transfer to retry or cancel. Managing the chip memory is not a simple task.

A technique to manage the memory is to use a boolean to indicate the success (or failure) of the entire transfer. A simple little class allocates and deallocates the memory according to whether the algorithm completes successfully. If it completes, the memory is fine and will be deallocated by the last part of the transfer. If any error occurs, success is not set to `true` and all the allocated memory is deallocated. This is shown below:

```
struct memory_watcher {
    memory_watcher(uint32 size, bool* success) :
    memory_(chip_allocate(size)), success_(success) {}
    ~memory_watcher()
      {if (! success_) chip_dealloc (memory_);}
    dut_pointer memory_;
private:
    bool* success_;
}
```

Now that we have a utility object to watch our memory allocations, we can code the algorithm without regard to handling memory:

```
//executes a control transfer, returns allocated memory
memory_list perform_usb_control_transfer(int id)
{
  bool success(false);
  memory_watcher request (STD_REQUEST_SIZE, &success);
  //set fields and send transaction
  memory_watcher ack (STD_ACK_NAK_SIZE, &success);
  if (! chip_perform_control_request (ack.memory_))
    return memory_list();//First error return

  memory_watcher data(STD_CONTROL_DATA_SIZE, &success);
  memory_watcher ack2(STD_ACK_NAK_SIZE, &success);
  if (! chip_perform_data_transfer(ack2.memory_))
    return memory_list(); //Second error return

  memory_watcher data (STD_STATUS_SIZE, &success);
  memory_watcher ack3 (STD_ACK_NAK_SIZE, &success);
  if (! chip_preform_data_ack_transfer(ack23memory_))
    return memory_list(); //Third error return

  success = true;
  return memory_list (request.memory_, data.memory_,
                      ack2.memory_, status.memory_,
                      ack3.memory_);
}
```

Okay, so this example is a bit long—but production code is this complex. It would have been much more complicated if we had to have many "if" tests or goto's to handle the memory allocation and deallocation.

Sometimes a thicket of logic can be made clarified by using little classes that clean up if an error occurs.

What's in a Name?

For some reason, class names in C++ tend to be more important than structure names in C or modules in Verilog. Maybe this is because in C++ we can enforce what operations are allowed in a class, so we tend to pay more attention to their names. At any rate, this section provides guidance on how to make the transitions between file, class, and instance when finding your way around a verification system. As we have said many times in this handbook, it's up to you and your team to decide what conventions to use.

Keeping class name the same as file name

A common convention is to have the class name be the same as the name of the header file that declares the class. For example, it is much easier to find a class or definition by using the Unix `find` command directly, rather than piping it to `egrep`.[3]

A corollary convention is to have only one class declaration per file. However, there are a few exceptions to this guideline. One is when there are small utility objects that are used only right where the main class is used. Another exception is when you are writing VIPs and it is simpler for a user to understand the interface as a monolithic entity. Note that, in some cases, the monolithic header file may just contain "`#includes`" of other header files.

> *Consider having a one-to-one relationship between class and file. Exceptions are where there are tiny helper classes and when a group of classes is more important than the individual classes.*

[3.] What could be easier than this? `find <path> -name "*.cpp" -exec egrep -l -i "your search text" {} -print`

Keeping class and instance names related

While C++ allows you to use any identifier for a typename and a variable, strive to make them as similar as possible. This seems like an obvious guideline, but we programmers are a lazy bunch. It is simple to miss changing an instance name when a class or enum is changed. It takes work and typing to keep names simple. (Appreciate that we essentially type for a living.)

Consequently, when an instance of a class is created, try to name the instance the same as the class. Sometimes, if there are several instances of a class in the same scope, a "_<n>", where n is an alphanumeric variable, can be appended to the name. The reason is that this provides a good mental link to back to the class definition, which specifies what can be done with this instance. A counter-example is when a class provides some generic behavior that can be used in many contexts. For example, a register class may provide generic reads and writes, as well as take in an address in the constructor. In this case it is the mnemonic of the address that is the best name for the instance.

Here is another counterexample, from a project the authors worked on:

```
ht_vip ht_drv; //hard to remember that ht_drv is a ht_vip
//Is pex known in the project? Is mon better than monitor?
pex_mon a_pex_mon;
```

Note that C++ allows identifiers and typenames to have the same string, as in my_class my_class, but the identifier shadows the class in some contexts and can be confusing.

Instance names should be readily traceable to their class name.

Coding with Style

Coding conventions can quickly become a "religious war," something that is not productive for a project, team, or individual. As a remedy, this section presents some style conventions that have proved to be useful. However, as with all the other sections in this handbook, the recommendations made are not intended to be a set of rules.

Adhering to a single style may improve clarity, but only if the entire system is coded by a single person (think "My style is the best!") But even in this case, one's style often evolves over time and adapts to the style of a team. In the general case, the industry definition of "good style" evolves as well.

Because of the evolution of what makes "good style," differences in style are essential for the learning process.

Proceeding with caution

In general, coding conventions slow down good coders, and do not necessarily increase the readability of the code created by poor coders.[4] Understandable code is understandable code, and vice versa, independent of the conventions used. The goal of a coding convention should be to increase communication among the team members.

For teams that feel the need for a "team style," a "guidebook" is usually a better idea than a required coding style. This guidebook should include guidelines, with reasoning following each guideline. In addition, counterarguments where the guideline may not be appropriate should also be provided. If the entire team does not agree on some guidelines, it is a good idea to include both the pro and the con arguments, so that locally appropriate decisions can be made—including allowing each team member to decide. The presence of the counterargument also provides a framework should some assumptions change.

[4.] "All generalizations are false, including this one." — Mark Twain

The goal of a coding convention should be to increase communication among the team members, not slow down the fast coders.

General syntax conventions

One guideline is to use all lowercase identifiers, with underscores and separators. Identifiers are all the nonreserved words of C++: your variables, class names, methods, data, and enums. The reasoning behind this convention is that there is less time spent thinking about how to type an identifier. An exception to this guideline would be if the team wants to capitalize three-letter acronyms (TLAs) and macros.

A counterargument to this approach is that it can create long names.

An alternative convention uses capitalization to indicate an identifier's scope, or class. For example, method names could begin with a capital letter, while data members begin with a lower-case letter. The reasoning behind this guideline is to encode the type information in the case of the identifiers.

Consistent naming conventions can be useful, but beware of dogma[5].

Syntactic sugar

There are a few *pseudo keywords* that have proven useful. Used consistently, they can help the reader of your code figure out your design intent. The counterargument is that they add unenforceable keywords to the language. A misuse of, or mistake in, using the pseudo keywords takes the reader down the wrong path. So, if they are used, they should be considered as similar to comments.

---

[5]. "A fanatic is one who can't change his mind and won't change the subject." (Sir Winston Churchill)

Here are the definitions:

```
#define override virtual
#define non_virtual
#define cached
#define uncached
```

Using override and non_virtual

When a method is declared, it is useful to know if it is being declared anew or is being *overridden* from a base class. The reason for this is that overrides are of interest because they hint at what needed to be changed from the base. New methods, by contrast, hint at what capabilities the inherited class adds. The override define is a way for the coder to identify the methods that are being overridden in the inherited class.

Also, methods are nonvirtual by default. However, the coder may have just forgotten to declare a method virtual. It is surprising how long this can go unnoticed in a design. To indicate that the designer intended to make a function nonvirtual, the non_virtual define is used.

Here is a brief example:

```
class tdm_generator : public byte_stream_generator {
  public:
    non_virtual void update_start_offset(uint32 off);
  protected:
    override void generate_byte_();
    virtual  void setup_next_frame_();
    static   void calculate_crc();
};
```

The above example shows an inherited class tdm_generator, which adds a new nonvirtual method called update_start_offset. This is important because you can be assured that this class has the only implementation.

The example then overrides the generate_byte_() method, changing, or perhaps just implementing, the interface behavior. The example adds a brand new virtual method, setup_next_frame_(), which—because it's virtual—is intended to be modified by subsequent inherited classes. The example also adds a static method that can be used only by itself and inherited classes.

With the `override` *and* `non_virtual` *defines, a class method would always contain a type of either* `non_virtual`, `virtual`, `override`, *or* `static`.

Using owner and cached

When a pointer is passed into a method, it's unclear what is intended. If it's passed in for efficiency, a `const&` should be used—so it cannot be just for efficiency. But what if the intent is to hand off ownership? In this case, the define `owner` can be used. This tells the reader that the method is assuming ownership (or shared ownership, for referenced counted data), and will be deleted when it is no longer needed.

The `owner` tag is also used in a class factory method, to indicate that the requirement to delete the created object is being passed to the caller.

Here is how `owner` could be used:

```
owner can_node* build (can_type t, const std::string&);
void send_frame(owner frame* f);
```

In the `build()` function, somehow the function creates a pointer to a `can_node`. The `owner` in the return type says that the caller is now responsible for the instance.

In the `send_frame()` function, the exact opposite occurs. The caller is now done with the frame, and the `send_frame()` function assumes ownership.

Another question to ask is, Will the method cache a copy of the pointer? The answer is especially relevant with respect to constructors, but it does apply anywhere a pointer is passed. In this case, the define `cached` is used. If there is a method that causes the class to forget about the pointer, the define `uncached` is used.

Here is a short example:

```
class driver {
  public:
    driver(cached configuration* c);
    void add_qos_metric(cached qos_metric*);
    void remove_qos_metric(uncached qos_metric*);
};
```

In this example the constructor is caching a pointer to the `configuration`. This is important to know, because configuration changes will immediately be seen by the `driver` class. In the `add_qos_metric()` method, the `qos_metric` instance is being cached. Again, this is good to know, because it implies that the pointer is shared among multiple instances (that is, it is a connection). The pointer passed into `remove_qos_metric()` is no longer known to the driver after the call.

> Using the `owner` and `cached` *defines, a pointer in a formal parameter would either be undecorated, or it would have the* `cached` *or* `owner` *tag if its scope exceeds the scope of the function or ownership that is being transferred.*

Identifying private and protected members

Another convention used is to identify protected and private members (data and methods) by using a trailing underscore. This allows one to know quickly whether the method is "internal." It also allows one to look at an algorithm in a method and separate the "internal state" from the method's parameters.

The counter argument is that a method name may become public as the project evolves. Because C++ uses scope rules first, and then access rules, you are guaranteed that no code is publicly using the method. However, this guarantee is not present if the name had to be changed. Consequently, changing the name may cause compile errors.

> *Identifying private and protected members helps others learn about a class.*

Using Templates

We've touched on the subject of templates in previous chapters. This section is another visit, this time talking about how templates can be used as coding techniques.

Standard templates for data management

Templates were a difficult addition to early versions of C++. Furthermore, the ramifications for templated code are huge. Because a main tenet of C++ is efficiency, there are many subtle ways to define and use templating that seriously affect code efficiency. Nevertheless, let's take a step back and define templating.

Templating is code that defines prototype functions, classes, or data with parameters to indicate the generic classes, data types, and (sometimes) methods. A template is then instantiated with specific classes and data types. This is sort of like using parameters in a Verilog module.

There are two main parts to templating: defining the template, and using the template. The difficulty is not symmetric, and it is far easier to use a well-defined template than to build one. For this reason, we'll start by describing some of the standard templates of C++ that are used in verification.

The C++ language comes with a standard set of general program templates. These templates are well-documented, efficient, and debugged, and are one of the things that make verification in C++ beneficial. The effort involved in creating the standard templates cannot be overstated.

By far, the most common templates used are `std::map`, `std::list`, `std::vector`, and `std::dequeue`. This is because a significant part of the work of verification involves keeping lists of expected data and mappings among key information and data (to send or receive) or the objects that handle the data (generators and checkers).

A `std::map` is an arbitrary mapping between one entity, called the *key*, to another entity. The fact that the `std::map` makes only a few requirements on the key[6] makes it very useful. Also, a map's underlying management classes change according to the number of elements in the map, so that accesses are fairly close to optimal.

For example, suppose you are writing a communications interface, with a single transactor handling a large number of phone calls. The phone call numbering is, by convention, a three tuple of <DS3, DS1, DS0>, where DS3 is a number between 1 and 12, DS1 is a number between 1 and 28, and DS0 is a number between 1 and 32. That's 10,752 connections!

Any number of the connections can be active, yet for all the calls the data go through one 8-bit interface. The BFM, therefore, must convert between some clock timing and a <DS3, DS1, DS0>, and then, from this phone-call identifier triplet, to a generator of call data.

Here are classes and a mapping that can be used to do that:

```
class phone_id {
public:
  teal::reg ds3(0,4); //a register that is 4 bits wide
  teal::reg ds1(0,5);
  teal::reg ds0(0,5);
};


//global function < for the phone id
bool operator< (const phone_id& lhs,
                const phone_id& rhs){
  return ((lhs.ds3 < rhs.ds3) && (lhs.ds1 < rhs.ds1) &&
          (lhs.ds0 < rhs.ds0));
}


class phone_id_bfm {
public:
  void start_call (const phone_id& c, call_voice* v) {
    assert(! active_calls[c]);
    active_calls_[c] = v;
  }
  void stop_call (const phone_id& c, call_voice* v) {
    assert(active_calls[c] == v);
    active_calls_[c] = 0;
```

---

6. It requires that the key be able to be ordered, by means of the binary operator< (). This is done automatically by the language for all integral types and some common classes, such as strings. Otherwise, you write an operator< () binary function.

```
    }
  private:
    void on_clock() {
      //determine DS3,1,0 from clock and start of frame
      voice* v = active_calls_(phone_id (current_ds3,
                              current_ds1, current_ds0));
      data_ = v ? v->get_next_data() : IDLE_CHAR;
    }
  }
  std::map<channel, cached call_voice*> active_calls_;
  teal::vreg data_; //mapped to the chip input
  }
```

In actual operation, one would determine (probably randomly) when and on what `phone_id` a call should be created. After the `voice` and `phone_id` are created, `start_call()` would be called. This would cause the BFM to inject data when that call's `phone_id` is referenced by the clock. Sometime later, `stop_call()` would be called and the `phone_id` would go back to the idle data pattern.

Another standard template that is useful is `std::deque`. This template manages an ordered list of generic data by means of the optimized insertion and removal of data from the ends of the list (back and front). (Examples of using a `std::deque` are shown throughout this handbook.)

The `std::deque` is an example of a C++ container template. There are many others, all with different runtime costs for insertion, deletion, and traversal. These traversal operations are called *iterations*, and the objects they use are called *iterators*. The templates have iterators that sweep forward and backward through all the standard container classes. Again, this lets the verification team concentrate on the domain-specific aspects, leaving the generic coding to C++ templates.

In addition to the standard container classes, C++ also provides a set of standard algorithms, such as `sort` and `find`. These work with the containers, and are tested, efficient, and standard.

The size and depth of the C++ template library is stunning. Most standard container types and algorithms can be found to be already implemented and debugged.

Writing templates for adaptable code

Although templates are powerful, be careful in building new ones and using templates outside the standard template library. They are tricky to get right (that is, make bug free), and are more difficult than standard code to understand. In the extreme, using templates creates two types of coders: the creators and the users.

Templates create hidden dependencies on the data types that are going to be used. These dependencies are not known until someone tries to instantiate a template with a data type that does not have the appropriate methods (or data members). This can create some surprising and spectacular error messages during compilation.

However, when the data type is simple and the algorithms are well-defined, templates can be the cleanest way to foster code adaptation.[7]

Let's look at an example. Consider the fact that every verification test must check the outputs of the chip. This is called *scoreboarding,* or *checking.* One might have a generic checker, but the data type changes with every verification component. Instead of downcasting (not very OOP), one can use templating. In this case, templating is also better than downcasting, because the different types of checkers are not interchangeable (in other words, the classes are not sibling classes). The intent of the coder is enforced.

Here is a possible templated checker:

```
template <class data_type>
class checker {
  public:
    checker(const std::string& name) : name_(name) {};
    void note_sent(const data_type& s)
      {sent_.push_back(s);}
    void check_received(const data_type& r) const {
      if (sent_.size() == 0) {
        std::cout << "checker " << name_
                  << "no data to check against" <<
          std::endl;
      }
```

[7.] An alternative method is just to use `void*` and let the user downcast. It is a judgment call which method is simpler to understand and use.

```
       else {
         data_type sent(sent_.front ());
         sent_.pop_front();
         if (sent != r) {
           std::cout << "checker " << name_
                     << "sent != received" << std::endl;
         }
       }
     }
   private:
     std::string name_;
     std::deque<data_type> sent_;
};
```

In order to use this checker, you must instantiate one with a specific data type, as shown below:

```
checker<teal::vreg> reg_stream_checker;
```

This creates an instance of the template, with a `teal::vreg` being substituted for `data_type`.

Sometimes you can make code adaptable by using a template.

Summary

In this chapter we looked at some of the techniques used to create our code "art." We talked about being careful with "if" tests; they are a necessary evil that can complicate the code. We introduced the concept of the factory function, useful in building inherited classes.

We offered the advice that you should code only what you need to know. We then introduced a variety of useful coding tricks and techniques that experienced coders use to solve programming problems, including, but not limited to, the following:

- Using reservable resources and mutex
- Using register fields instead of hard-coded integers
- Using data members (always carefully!)

- Using idioms to provide structure

- Using singletons for global services

- Using virtual protected methods, to separate interface from implementation

- Using cached booleans to indicate success

- Using naming and coding conventions to express intent and understandability

We also presented reasons for considering additional techniques, such as the following:

- When to use references instead of pointers, and why constant references are preferred

- When to use enums or integers, and when you should mix them

- Why automatic construction and destruction can lead to code clarity

- Why coherent class naming is a good thing, and why the names of classes, files, and instances should be related

- Why consistent style and syntax are a good thing—if they are applied with intelligence

- Why standard templates and algorithms are useful for efficient coding

So, this chapter covered a large number of techniques. Remember, you don't have to use all of these tricks all of the time, but they are here for reference when you need them.

For Further Reading

- There is a plethora of books devoted to coding in C++. While many people have their favorites, a few of these books are universally popular. One book that does stand out is *Effective C++: 50 Specific Ways to Improve Your Programs and Design* (2nd edition), by Scott Meyers. However, you might want to wait until you have a few years of C++ experience before using this.

Part IV: Examples (Putting It All Together)

This is what the rest of the book has built up to. Everything discussed earlier in this handbook is applied here to examples that better resemble the real world. This is still a book, so examples need to be relatively simple or they would be incomprehensible, but our goal with these examples is to show what real hardware verification with C++ looks like.

The examples here build on everything discussed so far. They use Truss as an open verification structure, and Teal as a connection gasket. They apply the OOP techniques discussed throughout the code.

The examples were not specifically chosen or coded to highlight the strengths of Truss, Teal, or C++. Rather, they were coded to resemble real-life projects as much as possible. Our goal is to show realistic examples and creative solutions. We hope you can pick up an idea or two by reading this. (The accompanying CD also provides a few open-source VIPs that can come in handy.)

Block Level Testing

I can give you a six-word formula for success:
Think things through—then follow through.

Sir Walter Scott

In many endeavors, follow-through is everything. From sports to parenting, it's not only what you say, but what you do that is important. This chapter is the first of the "follow-through" chapters.

We use all the tools, tips, and techniques from the rest of the handbook and apply them to something resembling a real-world example. This is the first complete example of what a test system using C++ might look like.

We look at a block-level verification system. Later, we'll adapt this same system to be used at the full-chip level.

Overview

This chapter covers a block-level verification effort as part of a large project. The goal is to verify a UART 16550 RTL block, written in Verilog. To do this, we will build an environment that will not only verify the block but also provide adaptable verification components for later project stages.

The example presented here will show all verification components needed to do UART 16550 verification as well as a fully randomized test. Several points of interest in the code will be highlighted throughout the chapter. (We present code in a slightly different form from the source code on the accompanying CD.) Sometimes, we merge a the interface and implementation of a class together, although they are separated in the source code. Also, we may abbreviate a class interface or some method's code to get straight to the point.

This chapter differs from the Truss tutorial chapter (in Part II) in that it focuses more on the middle layers of a verification system instead of on the flow. The middle layers are where managing the complexity of a verification system comes mostly into play.

If you want to look closer at execution order, it's recommended that you start by referring to the *Truss standard test algorithm* known as the "*dance*" in the Truss Basics chapter. Then, with the "dance" as a reference, divide and conquer by using a ends-in approach. In other words, take a closer look at both the top level `test.cpp` (and its related `test_components`) and the interface aggregrator, `testbench.cpp`. This will help show the overall structure and flow of the environment.

This chapter will talk about a few things. First, we set up the example with a theory of operation. This section highlights the overall environment and the interfaces that are used.

Then we look at several points of interest in the code. These points of interest cover code complexity problems of the middle layers in a verification system. We present these middle-layer techniques in their order of execution, by first looking at power-on reset, then at configuration and traffic generation, and then at checking.

Finally, we show how all the pieces are connected together through the `testbench.cpp` and `test.cpp`. This includes details on how channels, configuration objects, and interface layers are instantiated.

Theory of Operation

Many systems have at least one UART connection. This may be for diagnostics, software debugging, or general communication. For this reason, a single UART serves as a good first block-level example.

Here are the main components involved in the simulation:

UART Example: Objects and Connections

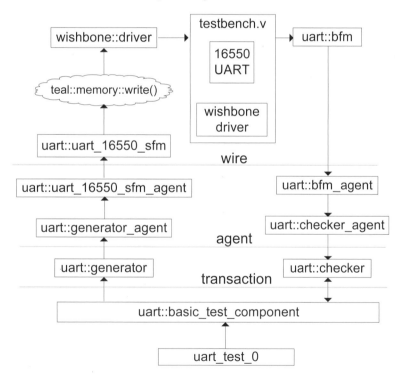

The UART Verilog was not developed by the authors. There are lots of implementations of UART cores available, and it was important to the authors that a known-to-be working UART model was used. For this example we choose to use an open-source design IP of a UART 16550 from Open Cores (see http://www.opencores.org).

This core follows the common register set of the 16550 UART, a popular UART implementation by National Semiconductor. It is so common, in fact, that software drivers for the UART 16550 are included with many Linux distributions. As with all design IP, this core has its own quirks that must be handled. We'll talk about this in the configuration section below.

The UART 16550 core used has two interfaces. One is the actual UART transmit and receive lines, and the other is a local bus to read and write registers in the UART block. In this case the local bus is a *wishbone* interface, a standardized local bus for many Open Cores models. Wishbone will be described in more detail in a later section.

Verification environment

The verification environment uses a UART BFM model to monitor the actual data transmitted and a wishbone driver model to read and write UART registers. For both interfaces, verification IP models, generators, BFMs, configuration objects, and checkers will be designed.

Looking at the Verilog side, the testbench environment is fairly straightforward. It is shown on the following page.

The UART module under test is called `uart_top`. It's instantiated in the `testbench.v`, file which also instantiates the wishbone Verilog driver and reference clocks. The rest is driven by C++ through Teal. The wishbone driver has a reset wire, which is used to reset the chip. These are the main components of the system.

UART Example: Wires and Objects

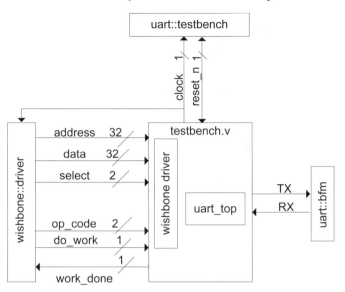

Verification IP

For this example, Verification IP (VIP) will be adapted or developed for the core's interfaces. VIPs are used to highlight how adaptable verification components can be developed and moved from one project to another. There is always work required when adapting an existing component to a new environment, but if the component is structured appropriately, the work can be minimal. VIP models will be provided for the UART and wishbone interface.

UART VIPs

For the UART 16550 verification system, the authors developed a generic UART BFM. We also developed what the authors call a *software functional model* (SFM). An SFM is a model of some protocol or common implementation that uses register access instead of wires for the chip connection. An example of this is the USB Open Host Controller Interface (OHCI). The specification defines what registers must exist and their meaning. It is similar to a BFM, except, instead of a bus, we are connecting to registers.

The generic UART BFM is specifically designed to be separate from the specifics of the 16550 protocol, so that the UART BFM can be used for any UART implementation.

The UART 16550 SFM, in turn, deals with all the registers for the 16550 protocol. This SFM acts like a software driver, in that the SFM programs the UART core's registers.

The UART 16550 SFM actually uses the wishbone BFM to set the registers of the Verilog core.

Wishbone VIP

The wishbone interface is the bus used to read and write the registers of the UART core. For the verification system, a wishbone BFM is used to read and write registers.

Instead of developing a new wishbone BFM from scratch, the authors decided to reuse a Verilog verification model provided with the UART 16550 core. This verification model included Verilog tasks for reading and writing registers over the wishbone interface.

These read and write tasks are wrapped into a C++ BFM. When a Verilog task is to be called (because of a higher-layer C++ call), the C++ side sets the appropriate wires and raises `do_work`. The Verilog side, in an `always` block, then calls the appropriate Verilog task or tasks, and when they return, raises the `work_done`. This signals the C++ side that the results from the Verilog driver are available.

Reusing existing verification models like this highlights how known-to be-working models can be integrated with a C++ verification environment. This technique is talked about more in the section on reading and writing registers.

The verification dance

The dance is the flow of events (or method calls) during simulation. It, of course, follows the "dance" talked about in the Truss Flow chapter. First, the chip is brought out of reset, then a configuration is chosen by means of randomization, and the UART core is configured (by means of register reads and writes). After this, a generator is asked to generate a group of data words for transmission. Because the UART protocol is bidirectional, both the ingress and egress sides have a generator and checker. After the data have been transmitted, the test waits for the checker to indicate that it received all the traffic. Then the test exits and a final status is printed.

Running the UART Example

Running the example is the same as for all tests that use Truss. However, before you run it, you'll need to set up some environment variables.

In the directory `/examples/single_uart/bin`, there is a setup script. Before you execute the script, make sure you have defined `BOOK_HOME`. Then source the setup script, and it will set `TRUSS_HOME` and `PROJECT_HOME` according to the `BOOK_HOME` variable.

Before you run the `truss` script, you must define the `SIMULATOR_HOME` environment variable. In addition, you must define `SIM`, for the simulator name and path to the install directory you are using. Type `truss -help` to see the currently supported list of simulators.

To run the example, type the following:

```
$TRUSS_HOME/bin/truss -test uart_test_0
```

There are many other options to `truss`, but this command will compile all the code and run the test. You should see a bunch of `c++` compiles and then the test will run.

Points of Interest

There are many points of interest in this first real example. There is a UART configuration object, used to pick the various settings for the protocol. There is a register interface, implemented as a Teal front-door memory bank. There are agent objects, used to connect the UART BFM and SFM to the generators and checkers.

So why are these objects important? In your verification system, you probably will have the equivalent objects for each of these points of interest. Understanding the reasons and trade-offs used to implement these object will provide you with a framework for deciding what is appropriate for your system.

Configuration

Most chip interfaces have a number of control registers to support configuration settings. These registers control the exact behavior of the interface; in other words, they describe what mode it is in. This might determine how the interface responds to interrupts or if the interface uses even or odd parity. With any real chip verification, there is a need to randomize this setup so that each configuration setting is tested.

By creating a *configuration class* for each interface, you create a centralized place that controls the randomization of each interface instance. This configurations class is not register biased. Rather, it contains protocol-generic features that are then mapped to registers by the specific interface implementation.

Why do this? For two reasons. First, you will most likely have a generic protocol side to the interface, which will operate outside of the chip. This IP will not have configuration registers (because it is not hardware), and it can execute the more-generic protocol configuration directly. Second, using a protocol-defined, but generic, class is a way to make the configuration adaptable to other implementations. Moving to a different core or even to a different chip with this same interface should not radically change the configuration class.

The configuration class is responsible for keeping track of all parameters of an interface, as well as for randomizing them into a "legal" configuration setting.

In our UART 16550 example there are two configuration classes: a generic class for the UART BFM, and a specialized UART 16550 class for the specific UART protocol we are testing. The specialized one inherits from the generic one.

Several techniques are used in the configuration objects to create adaptable code. These, or similar, techniques might be good to consider when you have to write your own interface. Here we will look closer at the different configuration classes, and highlight interesting areas.

VIP UART configuration class

The VIP provides a generic UART class that contains the largest legal configuration space. This is because it has been built to be valid for any UART core. The UART 16550 configuration object, by contrast, inherits from this interface, but actually limits the number of possibilities to what our core can support.

The UART configuration class is described below. It contains an `enum` definition for each parameter that describes the valid protocol domain, as well as a variable for each parameter.

The class also contains `randomize()` and `report()` methods to set up and print the status of the current setting.

Here are the interesting parts of the class definition:

```
namespace uart {
class configuration {
  public:
  configuration (const std::string& name);

  typedef enum {baud_rate_clock = 0x2000, rx, tx,
          dtr, dsr, cts, rts} signals;

  typedef enum {none=0, even, odd, mark, space} parity;
  /*rand*/ parity parity_;

  typedef enum {b_150 = 150, b_300 = 300, b_1200 = 1200,
```

```
                    b_2400 = 2400, b_4800 = 4800, b_9600 = 9600,
                    b_19200 = 19200, b_38400 = 38400,
                    b_57600 = 57600, b_115200 = 115200,
                    b_230400 = 230400, b_921600 = 921600
        } baud_rate;

        /*rand*/ baud_rate baud_rate_;
        /*rand*/ teal::uint8 data_size_;

        typedef enum {one=0, one_and_one_half,two} stop_bits;
        /*rand*/ stop_bits stop_bits_;

        virtual void randomize();
    }
```

The `randomize()` method is responsible for setting up each register to a legal value so that it can later be written to hardware or interpreted by a generic protocol VIP.

The reason each legal parameter is defined as an `enum` it to help show intent. This is important when writing adaptable code, and the idea with this UART model is that it should be able to be reasoned about and be adaptable to many different situations.

Because each parameter is an `enum`, some care must be taken for randomization. Let's look at this.

Randomization of enums

To randomize each `enum` in a simple way into a legal value, a couple of coding techniques have been used. Instead of picking a pure random value and comparing it against all possible legal values, the opposite approach has been taken.

Each legal register value is pushed into a standard container class and then is chosen by the generation of a random number that is based on the number of objects in the container. This requires a few more lines of code, but it eliminates the hassle of having to randomize an unconstrained integer over and over until a legal value is picked by coincidence. Also, because all the desired values are now in a single set, it's relatively simple to modify the distribution of the choosing of a value. (The separation between building a legal set and choosing a value can be exploited further, that is, by using subclassing, but this is beyond the scope of this handbook.)

To simplify this process further, operator overloading is used. By overloading the `operator++` for each enumeration defined, an instance of the enumeration can be incremented to its next legal value by a simple operation.

The `operator++` for the baud rate, as an example as defined in the `uart_configuration.cpp`, is shown here:

```
configuration::baud_rate operator++
        (configuration::baud_rate& m) {
  switch(m) {
    case configuration::b_150:
        return m = configuration::b_300;
    //...
    case configuration::b_460800:
        return m = configuration::b_921600;
    default: { truss_assert (0);};
  };
  truss_assert (0);
  return m = configuration::b_150;
}
```

As can be seen, each time the `baud_rate` is increased, the next legal `baud_rate` is automatically picked. This is used by the `randomize()` method to push each legal value into a container and to then pick a random `baud_rate` like this:

```
void uart::configuration::randomize()
{
  static baud_rate min_baud = static_cast<baud_rate>
      (dictionary::find(name + "_min_baud", 150));
  static baud_rate max_baud = static_cast<baud_rate>
      dictionary::find(name + "_max_baud", 921600));

  uint32 dead_man(0);
  std::vector<baud_rate> choices_b;
  for (baud_rate i(min_baud) ; i < max_baud; ++i) {
    choices_b.push_back(i);
    if (++dead_man >= 100000) break;
  }
```

```
truss_assert (dead_man < 100000) ;
choices_b.push_back(max_baud);
uint32 b_now(0);
RAND_RANGE(b_now, 0, choices_b.size() - 1);
baud_rate_ = choices_b[b_now];
//...
}
```

This code shows how the baud rate is randomly picked by the `random-ize()` method. The first grouping of interesting lines starts with `std::vector<baud_rate> choices_b`. This shows how all the possible baud rates are pushed into a vector. Note that the `++i` in the `for` loop uses the `operator++` as described above. This is how each legal value is pushed into an array. The power of C++ is harnessed for verification.

The second part is more straightforward. The lines starting with `uint32 dead_man(0)` show how a specific baud rate is picked, by randomizing over the number of objects in the container (that it, the size of the container) and then the chosen value is saved in the `baud_rate_` parameter.

This technique of defining `operator++` for each `enum` and using this to push each valid value into a vector is repeated for each register. Thus, at the end of the `randomize()` method, all registers are set to a random legal value.

UART 16550 configuration class

In this project a UART 16550 core IP is used. The UART 16550 is a common protocol, but our core puts a few restrictions on the legal UART register values. As shown below, we created a valid UART 16550 configuration by expanding upon the generic UART configuration class:

```
namespace uart {
class configuration_16550 : public configuration {
public:
  configuration_16550(const std::string& name) :
    configuration (name) {};

  typedef enum {interrupt, reference_clock} signals;

  virtual void randomize() {
```

```
//correct cases that our core cannot handle.
uart::configuration::randomize();
if ((stop_bits_ == configuration::two) &&
    (data_size_ == 5)) {
  stop_bits_ = configuration::one_and_one_half;
  log_ << teal_debug <<
    "Corrected stop bits from 2 down to 1.5"
    "data_size is 5)." << teal::endm;
}

if ((stop_bits_ ==configuration::one_and_one_half)
    && (data_size_ >= 6)) {
  stop_bits_ = configuration::two;
  log_ << teal_debug << "Move stop bits from 1.5"
    "up to 2 (data_size is 6, 7, or 8)." << endm;
}
}
};
};
```

The `configuration_16550` class inherits from the VIP configuration class in the same `namespace`. It overrides the `randomize()` method of the base `configuration` class. As shown in the implementation of the overloaded `randomize()` method, `configuration_16550` calls the base class `randomize()` method [see the `uart::configuration::randomize()` line] and then checks the actual values of a couple of registers.

If, for our core, illegal register combinations has been randomly chosen by the base class, `randomize()` corrects it. This is done to ensure that a legal UART 16550 configuration is picked.

Configuring the Chip

So how does an the actual chip get configured once a configuration object has been created and randomized for an interface? The configuration object represents the information a software driver would have to know to set the correct registers in actual chip.

In the Truss solution we follow this concept in the driver or BFM. A configuration object is known by all the particular drivers, BFMs, and monitors on an interface. This knowledge is necessary for the wire-layer objects to be able to drive and monitor the interface wires.

But how does the configuration get programmed to the actual chip? This is normally not done over the same interface. Programming the chip is normally done by one or a couple of major interfaces. For example, if a chip has an embedded processor, programming is mainly accomplished through the processor's external address and data wires. If the chip does not have a processor, this is accomplished through some external interface.

In our chip the wishbone interface is used to program the registers in the chip during the write-to-hardware phase of the "dance." The `write_to_hardware()` method of the `uart_16550_bfm` class doesn't access the hardware directly through its own wires. That would both complicate the code and made it harder to adapt. Instead, it uses the register defines on top of Teal's memory routines. The wishbone driver is hooked underneath these memory routines. Let's look at the technique of using Teal's memory access.

Register access

In order to be clear and to create adaptable code, the `uart_16550_bfm::write_to_hardware()` method uses register writes.

Here is the method:

```
void uart::uart_16550_bfm::write_to_hardware()
{
  register8 (data);
  data = 0;
  //...
  truss_assert(configuration_->data_size_ >= 5);
  truss_assert(configuration_->data_size_ <= 8);
  truss_field_put(data, data_size,
                  configuration_->data_size_ - 5);

  truss_field_put(data, access_clock_divide, 1);
  truss_reg_write(UART_REG_LC, data);
  register8( lc_save); lc_save = data;

  data = divisor;
  truss_reg_write (UART_REG_DL1, data);
  data = divisor >> 8;
  truss_reg_write (UART_REG_DL2, data);

  truss_field_put(lc_save, access_clock_divide, 0);
  truss_reg_write(UART_REG_LC, lc_save);
}
```

Notice that the code is using both truss_reg_write and truss_field_put. What are these functions? They are functions that can be used by both hardware and production software to access registers and fields within registers. There are defined (in /truss/inc/ truss_register.h) as follows:

```
void uart::uart_16550_bfm::write_to_hardware()
{
  register8 (data);
  data = 0;
  //...
  truss_assert(configuration_->data_size_ >= 5);
  truss_assert(configuration_->data_size_ <= 8);
  truss_field_put(data, data_size,
                  configuration_->data_size_ - 5);

  truss_field_put(data, access_clock_divide, 1);
  truss_reg_write(UART_REG_LC, data);
```

```
    register8( lc_save); lc_save = data;

    data = divisor;
    truss_reg_write (UART_REG_DL1, data);
    data = divisor >> 8;
    truss_reg_write (UART_REG_DL2, data);

    truss_field_put(lc_save, access_clock_divide, 0);
    truss_reg_write(UART_REG_LC, lc_save);
}
```

Why all this define trickery? The point is to abstract how the actual registers are accessed and manipulated. This way, both the hardware and software team can use the same macros. The authors are aware of, and have created, several fancier ways of accessing registers for verification. However, the authors believe that this mechanism has the appropriate level of simplicity and opens the door for reusing part of the verification code for software drivers or diagnostics.

Notice that the register addresses are defines. This is appropriate, although they could have been `const int` if the both teams decide to use C++ (as opposed to C, in which a lot of driver-level code is still written).

The field names are also defines, but they are named a specific way. This is because the `truss_field_put()` assumes a `_min` and a `_max` suffix to the field names. This was done to minimize the parameters into the macro.

For example, the following is used for the `data_size` field:

```
#define data_size_min 0
#define data_size_max 1
```

The implementation of `truss_reg_write` uses `teal::memory::write()`, which will find a memory bank mapped to that address and use it for the actual access.

Next we will look at how an actual address resolved to the wishbone interface.

The wishbone_memory_bank and wishbone_driver

Now we have seen how the UART 16550 SFM writes registers. But how does this get translated into accesses to the wishbone driver? Remember that Teal's memory routines use a look-up table to figure out which `memory_bank` object should handle the memory access. We'll just add a wishbone memory bank:

```
namespace wishbone {
class wishbone_driver;
class wishbone_memory_bank :
    public teal::memory::memory_bank {
  public:
    wishbone_memory_bank(const std::string n,
      const std::string& path, wishbone_driver* driver);

    teal::reg from_memory(teal::uint64 address);
    void to_memory (teal::uint64 a,
                    const teal::reg& d) {
      wishbone_driver_->write8 (address, value);
    };
};
```

The real work is done in the wishbone driver, although that, too, just calls down to a module in the Verilog. Here is how the write method of the driver works:

```
void wishbone::wishbone_driver::write8
          (teal::uint32 a, const teal::reg& d) {
  truss::mutex_sentry guard (mutex_);
  op_code_ = 0;
  address_ = a;
  truss_assert (d.bit_length () <= 8) ;
  //put the data on the right line
  switch (a % 4) {
    case 0: {select_ = 1; data_ = d; break;}
    case 1: {select_ = 2; data_ = d << 8; break;}
    case 2: {select_ = 4; data_ = d << 16; break;}
    case 3: {select_ = 8; data_ = d << 24; break;}
  }
  do_work_ = 1;              //signal to verilog
  teal::at (posedge (work_done_)); //wait for ack
```

```
}
```

By setting do_work_ to 1, we notify the Verilog of a pending transaction. By waiting for work_done_ to be 1, we cause the C++ code to wait until the Verilog half of the driver signals that the transaction completed.

The Verilog code is not really interesting, as it, in turn, just calls tasks in a module called wb_mast. This module is part of the Open Cores code. All these files are in the directory /verification/vip/wishbone.

This technique of adapting existing Verilog tasks is a good way to leverage working, debugged Verilog code. There is no need to throw it away, nor any need to rewrite it into C++.

Traffic Generation

Now that we have the chip all configured, we need to send traffic through it. The UART VIP code contained a basic generator, whose interface is shown below:

```cpp
namespace uart {
typedef truss:channel_put<block> channel_put;
class basic_generator {
  public:
    basic_generator(const std::string& n,
                    uart::channel_put* t,
                    const teal::uint8* word_size) ;

    //send one block of words to the uart bfm,
    //hold off sending the block by delay
    void send_block(teal::uint32 block_size,
                    teal::uint32 bit_delay);
  protected:
    void send_block_(teal::uint32 block_size,
                     teal::uint32 bit_delay) = 0;
  };
};
```

The send_block() method creates a block of data, with a specific block delay and then calls the connection virtual method send_block_(). The data word size is fixed because the configuration has been randomized previously.

The `send_block_()` is a pure virtual method and is used as the agent connection to the BFM or SFM. The agents are discussed next.

The generator_agent and uart_bfm_agent classes

Now that the generator is generating traffic, we have to connect it to the BFM or SFM. There are as many ways to do this are there are stars in the sky. The authors have chosen to have the connection agents use channels.

```cpp
namespace uart {
typedef truss::channel_put<block> channel_put;
class generator_agent : public generator {
  public:
    generator_agent(const std::string& n,
                    uart::channel_put* t,
                    const teal::uint8* word_size) ;
  protected:
    virtual void send_block_(const block&){
        out_->put(the_block);
    }
  private:
    generator_agent(const generator_agent&);
    void operator=(const generator_agent&);
    channel_put* out_;
};
};
```

This class does not contain much code. Remember, the purpose of this class within the Truss framework is to enable a connection-policy decision, so the code size is secondary. That said, smaller is better, and this example relies on the channel to do most of the work. It simply puts the data to be sent into a channel.

Let's take a look at the other half, the connection to the BFM or SFM. We'll only show the BFM as the SFM is strikingly similar. Of course, these agents should use a channel as well, because the testbench connects instances of these two classes together.

This class is shown below:

```
namespace uart {
typedef truss::channel_put<block> received_from_wire;
typedef truss::channel_get<block> to_be_transmitted;
class bfm_agent : public bfm {
  public:
    bfm_agent(const std::string& name,
        truss::port <configuration::signals>::pins port,
            const configuration* c,
            to_be_transmitted* to_be_transmitted,
            received_from_wire* received_from_wire,
            teal::uint64 clock_frequency);

  protected:
    virtual void receive_completed_(const word&) {
      block current_rx = block (0);
      current_rx.add_word(current_rx_word);
      received_from_wire_->put(current_rx);
    }

  private:
    void do_tx_thread();
      for (;;) {
        block current_tx = to_be_transmitted_->get();
        if (current_tx.block_delay_) {
          pause_(one_bit_ * current_tx.block_delay_);  }
        }
        for (std::deque<word>::iterator
            it (current_tx.words_.begin());
            (it != current_tx.words_.end()); ++it) {
          send_word(*it);
        }
      }
    }
};
};
```

There are a few point of interest here. Since UART is a bidirectional protocol, there are two channels. One channel is used to connect to the checker agent, and the other channel is used to connect to the generator agent.

Another effect of the UART being a bidirectional protocol is that there are two methods, one to support each channel. One method is the connection technique of overriding of a pure virtual method, in this case, `receive_completed_()`.

The other channel-supporting method is `do_tx_thread()`. As you can probably guess, this method runs in a separate thread of execution. This method first delays the appropriate amount. It then takes the block of data words and sends them, one at a time, to the UART BFM.

There is one more point to make before we move on. A chip might have several ways to drive an interface, such as register, FIFO, or DMA. One would probably write corresponding SFMs and SFM agents.

In general, the agents implement a connection policy by overriding the pure virtual method in the base class. In this example, we used a channel policy.

The Checker

Now that we have the transmit side connected, let's take a look at the checking side. We have already done half the work. The agents will place any received data into a channel. We just need to create the checker agent to connect the channel to the checker, as follows:

```
namespace uart {
typedef truss::channel_get<uart::block> inputs;

class checker_agent : public checker {
  public:
    checker_agent(const std::string& name,
                  inputs* expected, inputs* actual);

  protected:
    virtual void get_expected_(uart::block*) {
      truss_assert(the_block)
      *the_block = expected_->get();
    }

    virtual void get_actual_(uart::block*) {
      truss_assert(the_block)
      *the_block = actual_->get();
```

```
        }

        virtual bool more_() { return expected_->size();}

        inputs* expected_;
        inputs* actual_;

    private:
      checker_agent(const checker_agent&);
      void operator=(const checker_agent&);
    };
  };
```

The checker agent is providing the connection policy for the checker. As we have used a channel for the connection, channels are used here. The expected channel comes from the generator (through a tee, or tap, described in the next section), and the actual channel is the received channel in the BFM.

Detailed BFM Agent Connections

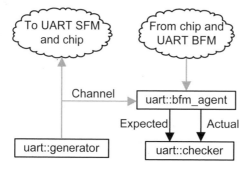

The checker_agent provides the checker with three things: the expected, the actual, and whether or not there is more checking to do.

Checking the data

Let's take a look at the checker. It's a little complex because the checker must handle the fact that the expected and actual block sizes may be different. This is expected in a real system, for several reasons. One is that a DMA- or FIFO-based receive will "clump up" the received data, depending on how the chip was set up (the specific FIFO interrupt trigger points, DMA block sizes, and so on). Another reason is because a transmission may have to be broken up into segments by the protocol.

Here is the key algorithm in the checker. In the text below, we have removed most of the received data code. This is because it is identical to the expected code. Also, the code has been simplified just a little, but the essence is still the same. (The actual code is on the accompanying CD.)

```
void uart::checker::perform_checking_() {
  uart::block current_tx_block;
  std::deque<uart::word>::iterator current_tx;

  for (;;) {
    if (current_tx == current_tx_block.words_.end()) {
      get_expected_(&current_tx_block);
      current_tx = current_tx_block.words_.begin();
    }

    if (*current_tx != *current_rx) {
      //... Long error print here!
    }

    ++current_tx;
    if ((current_tx == current_tx_block.words_.end())
        && (!more_()))
      done_.signal(); }
  }
}
```

The algorithm compares the data words and relies on STL iterators to move through the block of words. When the iterators hit the end, a new block of data is pulled from the agent. The algorithm also uses the uart::word::operator==() to decide how to compare the block elements. The power of C++ is harnessed for verification.

If the agent indicates that there are no more blocks, we signal an event. This event is used by the wait_for_completion() code, in a stunning display of software engineering:

```
void uart::checker::wait_for_completion()
{
  done_.wait();
}
```

The test, by way of the test component, is waiting for this checker's `wait_for_completion()` to return, signifying that the test is done.

We have made it through the bulk of the code. The next thing we need to talk about is how the objects are built and hooked together. Then, we need to talk briefly about the test component and an example test.

Connecting It All Together

The previous sections have discussed the interface components at the various layers of abstraction. Now its time to put them together. The first place we'll start is the testbench, as it will create the instances and connect them by means of channels. Then we'll take a brief look at the test component, which exercises the ingress or egress flow of traffic. Finally, we'll look at a basic test to send data in both directions.

The testbench

The testbench is responsible for building the interface components and bringing the chip out of reset. We will not discuss bringing the chip out of reset, as it is pretty much the same as in the tutorial. Building the components, however, is something new.

The components are built in the testbench constructor. We will look at the constructor in stages, as several different things are happening. First, let's look at some naming conventions that will be used in the testbench.

As the UART protocol is bidirectional there is a name for the traffic flow in each direction. We will use the industry standard terms of *egress* for traffic originating from the chip and flowing outward, and *ingress* for traffic flowing inward.

Building the channels

Did you notice that the agents' constructors need only a `channel_get` or a `channel_put`? The idea is that each side of the agent class only needs to get items out of or put items into a channel, not both.

Now, here in the testbench, we build a "real" channel, which is the merging of the two abstract channel base classes, `channel_get` and `channel_put`, as shown below:

```
uart::channel* software_egress =
    new uart::channel ("Software egress");
uart::channel* software_egress_tap =
    new uart::channel ("Software egress tap");
software_egress->add_listner (software_egress_tap);
channel* software_ingress =
    new uart::channel("Software ingress");

channel* protocol_ingress =
    new uart::channel("protocol ingress");
channel* protocol_ingress_tap =
    new uart::channel ("protocol ingress tap");
protocol_ingress->add_listener(protocol_ingress_tap);
channel* protocol_egress =
    new uart::channel("protocol egress");
```

Notice the `add_listener()` call, which connects the `out()` method of one channel to an arbitrary number of other channels. We use it here to give the checker a copy of the generated data.

Building the configuration and interface port

After the channels have been built, there are two more things we need to build before creating the models. They are the configuration and the path to the pins.

Building the configuration is straightforward:

```
uart_configuration = new uart::configuration_16550(n);
```

Building the path to the pins can be done in many ways. In this example, we just concatenate the top path with the pins that we need:

```
port <uart::configuration::signals>::pins
        protocol_port;
protocol_port[uart::configuration::baud_rate_clock] =
        top_ + ".BAUD_RATE_CLOCK";
//...
protocol_port[uart::configuration::cts] = top_ + ".RTS";
```

```
protocol_port[uart::configuration::rts] = top_ + ".CTS";
protocol_port[uart::configuration::rx] = top_ + ".TX";
protocol_port[uart::configuration::tx] = top_ + ".RX";
```

Notice that the ports are templated on the configuration enumeration signals. Truss provides a small templated typedef that maps an enum to a string. There are a few benefits to this technique. One benefit is that the intent of passing the pins is made clear (as opposed to a list of strings). Another benefit is that the mapping between pin and enum is explicit and well-bounded. Finally, because the concept is templated, you can adapt the technique to your set of pins.

Building the interface-layer objects

Now we are ready to build the objects of the interface layer. This code is shown below:

```
uart_protocol_bfm = new uart::bfm_agent(
    "uart Protocol", protocol_port,uart_configuration,
    protocol_ingress, protocol_egress,
    UART_CLOCK_FREQUENCY);
uart_software_bfm  = new uart::uart_16550_agent(
    "16550 uart",software_port,uart_configuration,
    program_egress, program_ingress,
    UART_CLOCK_FREQUENCY);

    //build and hook up the ingress and egress stimulus and
scoreboards of the interface
uart_egress_generator = new uart::generator_agent (
    "egress_generator", software_egress,
    &uart_configuration->data_size_);
uart_ingress_generator = new uart::generator_agent (
    "ingress_generator", protocol_ingress,
    &uart_configuration->data_size_);

uart_ingress_checker = new uart::checker_agent(
    "ingress checker", protocol_ingress_tap,
    software_ingress);
uart_egress_checker = new uart::checker_agent(
    "egress checker", software_egress_tap,
    protocol_egress);
```

The generator and checker are used for both sides of the interface. This is appropriate, because the generator and checker should have no idea of the connection policy or actual implementation details of the interface.

We are almost done with building all the lower-layer objects. We just need to create the register access code.

The wishbone objects

Building the wishbone objects is just a matter of building a driver and memory bank, and then mapping the memory bank to an address range, as follows:

```
port <configuration::signals>::pins  wb;
//...
wb [wishbone::configuration::address] =
  top_ + ".wishbone_driver.address";
wb [wishbone::configuration::data] =
  top_ + ".wishbone_driver.data";
wb [wishbone::configuration::do_work] =
   top_ + ".wishbone_driver.do_work";
wb [wishbone::configuration::work_done] =
   top_ + ".wishbone_driver.work_done";

wishbone_driver_ =
  new wishbone::wishbone_driver("WB", wb);
teal::memory::add_memory_bank(
  new wishbone::wishbone_memory_bank
    ("Wishbone", "main_bus", wishbone_driver_));
teal::memory::add_map ("main_bus",
        uart_registers_first, uart_registers_last);
```

The wishbone_driver is built and handed to the wishbone_memory_bank, which caches the pointer. Then, the wishbone_memory_bank is added into the Teal memory system. Finally, this newly added bank is mapped to the first through the last register address of the UART 16550 interface of our chip.

That's it! From this point in the code and onward, any memory::write()s or memory::read()s to that address range will go through the wishbone_memory_bank and then to the driver.

Whew, that was a lot of code! However, building all the components of a testbench is a large job. We'll now move up a level, looking at the test component and then the test.

The test component

Compared to the testbench, the test component is simple. The testbench pretty much just forwards its dance calls to the appropriate generator, model, or checker, as follows:

```
namespace uart {
class basic_test_component :
    public truss::test_component {
public:
  basic_test_component(const std::string&, generator*,
          verification_component* model, checker* c);

  virtual void randomize(); //shown in next section
  virtual void time_zero_setup()
      {model_->time_zero_setup ();};
  virtual void out_of_reset(reset r)
      {model_->out_of_reset (r);};
  virtual void write_to_hardware()
      {model_->write_to_hardware();}

protected:
  virtual void start_components_()
      {model_->start(); checker_->start();}
  virtual void generate()
   {generator_->send_block (block_size_,block_delay_);}
  virtual void wait_for_completion_()
      {checker_->wait_for_completion();}

  /*rand*/ teal::uint32 block_size_;
  /*rand*/ teal::uint32 block_delay_;
  };
};
```

We won't go over the code above in detail; just take a look and notice that most of the methods are one-line calls to the appropriate interface-layer component.

The last two lines are interesting. They are the random variables that are used by the `generate()` method to create random data. These are the variables that will be controlled by the test (as well as by configuration variables).

The uart_basic_test_component::randomize() method

The `generate()` method is where the test component sends traffic through the interface. It sends only one group of data, but that group length can be any size. The next chapter shows how this method can be called repeatedly.

The `generate()` method only does what it is told. The `randomize()` method is responsible for choosing the appropriate block length and delay for the block. Why do we seperate these two related methods? Because you may want different constraints and distributions for the random parameters. Note the following:

```
void uart::basic_test_component::randomize()
{
static uint8 min_words = dictionary::find(name +
      "_min_num_words", 2);
static uint8 max_words = dictionary::find(name +
      "_max_num_words", 4);
static uint8 min_bit_delay = dictionary::find(name +
      "_min_block_delay", 0);
static uint8 max_bit_delay = dictionary::find(name +
      "_max_block_delay", 10);

block_size_ = get_block_size(min_words, max_words);
block_delay_ =
         get_bit_delay(min_bit_delay,max_bit_delay);
}
```

Teal's dictionary is used to see if any high-level code (such as a test) has overridden the parameters. Then, two local simple static functions are used to generate the values, subject to the minimum and maximum specified. Here is the `get_bit_delay()` function:

```
namespace {
  uint8 get_bit_delay (uint8 min_v, uint8 max_v) {
```

```
        uint8 returned;
        RAND_RANGE (returned, min_v, max_v);
        return returned;
    }
}
```

That's all there is to the test component. Once a test creates one and follows the standard Truss dance, traffic will be sent and checked through the UART interface!

Now let's take a look at the test.

The basic data test

The only top-most component that we have not talked about is the test. The test, like the test component, is straightforward. That is as we expect, because the top-most layers should be obvious.

The test is fairly unremarkable. Here is an abbreviated look at its interface:

```
class uart_test_0 : public truss::test_base {
 public:
   uart_test_0(testbench* tb, truss::watchdog* wd,
              const std::string& name);
   //... All the usual dance methods, for example...
   virtual void write_to_hardware() {
     uart_test_component_egress_->write_to_hardware();
     uart_test_component_ingress_->write_to_hardware();
   }
 private:
   testbench* testbench_;
   uart::basic_test_component*
                             uart_test_component_ingress_;
   uart::basic_test_component*
                             uart_test_component_egress_;
};
```

The test builds two test components, one for inbound traffic and one for outbound traffic. For each method, it just calls the same named method on each component.

The authors realize that this can seem tedious, but at least you have all the control. If you need to do some special pre- or postprocessing, it's a simple matter to add it. If you don't want to call all the test components' methods all the time, just leave it out. It there is a specific order you need, or you need some extra communication between the test and the test components, you can just add them.

One alternative, which the authors have worked on, is to have a global sequencer. This is almost always a mistake, in that it makes the test writer's job harder. Remember the guideline—that "tedious and obvious" is preferable to "less code and hidden."

The interesting part of the test is in the constructor, as shown below:

```
uart_test_0::uart_test_0(testbench* tb,
                         const std::string& n) :
test_base(n), testbench_(tb),
uart_test_component_ingress_(new
  uart::basic_test_component("uart_ingress",
    tb->uart_ingress_generator,tb->uart_program_bfm,
    tb->uart_ingress_checker)),

  uart_test_component_egress_ (new
   uart::basic_test_component("uart_egress",
    tb->uart_egress_generator, tb->uart_protocol_bfm,
    tb->uart_egress_checker))
{
//add configuration default constraints
teal::dictionary::put(
  tb->uart_configuration->name + "_min_baud", "4800",
  teal::dictionary::default_only);
teal::dictionary::put (
  tb->uart_configuration->name + "_min_data_size", "5",
  teal::dictionary::default_only);
teal::dictionary::put (
  tb->uart_configuration->name + "_man_data_size", "8",
  teal::dictionary::default_only);

  //add generator default constraints
teal::dictionary::put(
 tb->uart_egress_generator->name+"_min_word_delay", "1"
 teal::dictionary::default_only);
teal::dictionary::put(
  tb->uart_egress_generator->name+"_max_word_delay", "3"
  teal::dictionary::default_only);
```

```
//...
}
```

This code does two things. First, it creates and wires up the ingress and egress test components. Second, the constructor adds some parameter values to guide the configuration selected and the amount of data to be sent.

That it! We've made it through the first real-world test system!

More Tests

While the test in the example is sufficient for most of the "normal" cases, there are still several things we should do to test the core fully. Besides the additional features of the core, like loopback and FIFO depth triggering, there are a range of error tests to be performed.

For example, one can test parity errors or stop bits, or perhaps the sampling algorithm for the data bits.

There are also the external control pins, such DTR, DSR, and so on, that should be exercised.

All of these tests, which must be written and performed, are beyond the scope of this handbook.

Summary

This chapter ties together the last couple of hundred pages or so. We built a verification system to unit-test a UART.

A configuration convention was covered. Truss does not address chip configuration, because this is chip- and feature-specific. We did show how the Teal dictionary can be used to get and set parameters globally.

An interesting part of configuring the chip was using the Truss register defines with the Teal memory space. This provided a generic register interface that could be mapped to any memory bank. In our case, we adapted a wishbone Verilog model.

The policy of channels was selected to connect the transaction level classes with the wire-layer ones. We used the Truss templated channel.

Checking the data was a little complicated, because the packets to be checked were possibly a different size from when they were generated.

The test component, testbench, and test were described, with an emphasis on the testbench constructor. This was where all the interface objects were created and the channels connected.

Chip Level Testing

CHAPTER 15

And will you succeed? Yes indeed, yes indeed!
Ninety-eight and three-quarters percent
guaranteed!

Dr. Seuss

Testing at the block level is common. Testing at the system level is necessary. As you probably know, it's the system level interaction between the various blocks that must be tested. This system level interaction is the focus of this chapter.

This chapter presents three main concepts:

- The chip now has four UART interfaces.

- We develop three tests, showing a progression from getting the interfaces running to a generic test with irritators.

- We can adapt the original block-level test to be used in the system.

Overview

This chapter highlights Truss irritators. We'll adapt the UART block-level testbench to a system-level testbench that has four UARTS. One of these UARTS will be randomly chosen to be the focus of the test, while the other three will serve as background traffic irritators. While this chapter uses UARTs for the irritators, the idea is generic.

Theory of Operation

This verification system builds upon the block-level UART system. We will adapt the components developed in the last chapter, and add a few new tests. These tests will show how irritators are used.

Here are the main components involved in the simulation:

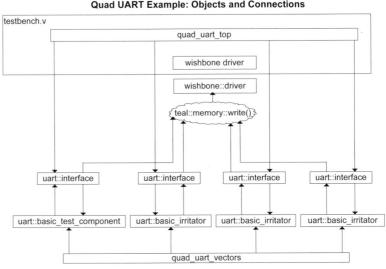

Quad UART Example: Objects and Connections

One difference from the block-level system is that the testbench now does not directly build each of the four UART interface's objects (such as the generator, checkers, agents, and so on). This is left to the

Hardware Verification with C++

`uart::interface` class, which is called by the testbench. Another difference is that three of the interfaces are exercised by irritator objects.

As with all good projects, this project steals (er, adapts) code from the block-level UART. The UART BFM, generators, and checkers are reused without modification. We modify the UART 16550 SFM by adding an integer ID, which is used to form the specific UART address.

The UART test component is also reused directly. In addition, we inherit from the test component to create an irritator class. So, we added only a few lines of code to the block-level test.

We will develop four tests. One is the previous chapter's block-level test, with modifications to select one of the four UARTs. The other three tests exercise all four UARTs at once and show the test development progression from a simple group of test components to a single test component and a list of irritators. Because the irritators are a common base class, this final test can be used as model for random tests in the Truss environment.

Verification environment

Looking at the verification environment, we see that it is very similar to that for the single UART, with no significant differences. We still have the wishbone driver to access the registers, this time mapped to a larger region. We still have all the main players from the block-level UART.

Looking at the HDL side, the testbench environment is fairly straight-forward. The only different module, `quad_uart_top`, instantiates four UARTS and maps their write enable and read select according to the upper address bits. The testbench environment is shown below.

Quad UART Example: Wires and Objects

Running the UART Example

There are four tests in the example. Running the example is the same as for all tests that use Truss. To run a test, select it and type one of the following:

```
$TRUSS_HOME/bin/truss -test uart_test_0
$TRUSS_HOME/bin/truss -test quad_test_components
$TRUSS_HOME/bin/truss -test quad_uart_irritators
$TRUSS_HOME/bin/truss -test quad_uart_vectors
```

The quad_test_components Test

The quad_test_components test is the first test for the chip that the authors wrote. We were just making sure that all the UARTs could be addressed. This test chooses four different random configurations and sends a random-length block (with random block delays) to each ingress and egress channel. Here is the constructor part of the test:

```
quad_test_components::quad_test_components(
        testbench* tb,
        truss::watchdog* w, const std::string& n) :
        truss::test_base(n, w), testbench_(tb){
truss_assert(number_of_uarts >= 2);
for (teal::uint32 i(0); i < number_of_uarts; ++i) {
  std::ostringstream o; o << i;
  std::string id = o.str();
  uart::interface* if = tb->uart_interface[i];

  uart_test_component_ingress_[i] = new
    basic_test_component("uart_ingress "+ id,
            if->uart_ingress_generator,
            if->uart_program_sfm,
            if->uart_ingress_checker);

  do_generator(
        if->uart_ingress_generator->name);
  uart_test_component_ingress_[i]->randomize ();

  uart_test_component_egress_[i] = new
    basic_test_component("uart_egress " + id,
            if->uart_egress_generator,
            if->uart_protocol_bfm,
            if->uart_egress_checker);

  do_generator (
    if->uart_egress_generator->name);
  uart_test_component_egress_[i]->randomize ();

  do_configuration (
    if->uart_configuration->name);
  }
}
```

This example scales up nicely from the block-level test. The rest of the test's methods are fairly boilerplate, and do not need any special behavior.

The quad_uart_irritators Test

The quad_uart_irritators test is the second test for the chip that the authors wrote. In this test we have randomly selected one UART as the focus of the test (that is, by using a test_component), and we have an array of three UART irritators. What are these uart_irritators? Let's take a look at the class.

UART irritator class

Remember that a Truss irritator is a way to adapt a test component to run as background traffic. The test component's run_traffic() method will be called repeatedly until the test decides its time to stop. Here is the interface for the UART test irritator:

```
namespace uart {
  class basic_irritator : public truss::irritator,
                          public basic_test_component {
  public:
    basic_irritator(const std::string& n,
                    uart::generator* g,
                    truss::verification_component* b,
                    uart::checker* c):
      truss::thread(n), truss::irritator(n),
      basic_test_component(n,g,b,c),test_component (n){}

    virtual ~basic_irritator() {}
    virtual void start() {irritator::start();}
    virtual void stop() {irritator::stop();}
    //... Shown below.
  private:
    basic_irritator(const basic_irritator&);
    basic_irritator operator=(const basic_irritator&);
  }
};
```

The uart::basic_irritator is both a truss:irritator and a basic_test_component. This makes sense, as the interface will follow the Truss irritator model and the implementation will use the uart_basic_test_component, which is unchanged from the block-level test. This is a really great thing about Truss: the same test component can be reused from the block level to the system level. Also, because a testbench is not mentioned anywhere in the test component, the test component can be moved to different projects easily. This is not accidental, but rather is a direct result of the layered approach talked about earlier in this handbook.

Let's take a look at the rest of the methods.

```
virtual void report(const std::string prefix) const
  {basic_test_component::report(prefix);}
virtual void time_zero_setup()
  {basic_test_component::time_zero_setup();}
virtual void out_of_reset(reset r)
  {basic_test_component::out_of_reset(r);}
virtual void write_to_hardware()
  {basic_test_component::write_to_hardware();}
virtual void wait_for_completion()
  {basic_test_component::wait_for_completion();}

 protected:
 virtual void randomize()
   {basic_test_component::randomize();}
 virtual void wait_for_completion_()
   {basic_test_component::wait_for_completion_();}
 virtual void start_components_()
   {basic_test_component::start_components_();}
 virtual void generate()
    {basic_test_component::generate();}

 virtual void inter_generate_gap()
   {checker_->wait_actual_check();}
```

Notice that all the methods (save one) just call their test_component base method. This is a standard form, tedious indeed, but it gives the coder the ability to add special code if needed. Again, this is the tedious but obvious guideline coming into play.

There is one method, inter_generate_gap(), that is not just calling the test_component's method. This is because this method is specific to an irritator. In our case, we know that the checker is derived from the Truss checker, and so has a method to wait until expected or actual data are checked. This is an appropriate throttling method for our irritator.

As coded here, this method pauses the generation until a data packet is checked. We could have done fancier things, such as have an initial number of packets in play, or change the delay depending on the actual data bytes generated.

That's it for the irritator! In about 20 lines or so of code, we have added the ability to use any UART interface as background traffic. Furthermore, the interface is that of a generic irritator, able to be plugged into any test that has a list of irritators. This is shown in quad_test_irritators. Let's take a look at the test.

The test

This test is the first one to use irritators. Of course, we also have a test component that is the focus of the test. This test is a little confusing, in that we use a UART for both the test component and the irritators. Nevertheless, this is what we have to test. Here are the interesting parts of the test's interface:

```
const uint32 irritator_count = number_of_uarts - 1;

class quad_uart_irritators : public truss::test_base {
 public:
   quad_uart_irritators(testbench* tb, watchdog* w,
                        const std::string& name);
   //...
 private:
    testbench* testbench_;
    //The focus of the test.
    uart::basic_test_component* uart_ingress_;
    uart::basic_test_component* uart_egress_;

    //The background traffic components.
    uart::basic_irritator*
       uart_irritator_ingress_[irritator_count];
    uart::basic_irritator*
       uart_irritator_egress_[irritator_count];
};
```

This test uses a fixed array of irritators. In this test they are explicitly called out as uart::basic_irritator. This use of a specific irritator type will be made more generic in the next test. Let's take a look at the implementation of the constructor:

```
quad_uart_irritators::quad_uart_irritators(testbench* tb,
truss::watchdog* w, const std::string& n) :
   truss::test_base(n, w), testbench_(tb)
{
   truss_assert(number_of_uarts >= 2);
}
```

Where did all the code to initialize the test components go? That code is moved into the `randomize()` method, because now the test has some random behavior. In this case, randomization determines which UART interface to pick for the test_component. Here is the `randomize()` method:

```
void quad_uart_irritators::randomize() {
//First, for the  main point of the test...
   static uint32 min_index =
      dictionary::find(name_ + "_min_uart_index", 0);
   static uint32 max_index =
      dictionary::find(name_ + "_max_uart_index", 0);

   uint32 selected = get_index(min_index, max_index);
   log_ << teal_info << "Selected uart "
        << selected << endm;
   std::ostringstream o; o << selected;
   std::string id = o.str();
   uart::interface* if = tb_->uart_interface[selected];

   do_configuration(if->uart_configuration->name);

   uart_test_component_ingress_ =
      new basic_test_component (
         "uart_test_component_ingress " + id,
         if->uart_ingress_generator,
         if->uart_program_sfm,
         if->uart_ingress_checker);
   do_generator (if->uart_ingress_generator->name);
   uart_test_component_ingress_->randomize();

   uart_test_component_egress_  =
      new basic_test_component (
```

```
            "uart_test_component_egress " + id,
            if->uart_egress_generator,
            if->uart_protocol_bfm,
            if->uart_egress_checker);
        do_generator (if->uart_egress_generator->name);
        uart_test_component_egress_->randomize();


    //now for the irritators...
    teal::uint32 count = 0;
    for (teal::uint32 i(0); i < number_of_uarts; ++i) {
      o << i; id = o.str();
      if = tb_->uart_interface[i];

      truss_assert(count < irritator_count);
      if (i != selected) {
        uart_irritator_ingress_[count] =
          new basic_irritator ("irritator_ingress "+ id,
                  if->uart_ingress_generator,
                  if->uart_program_sfm,
                  if->uart_ingress_checker);
        uart_irritator_egress_[count]  =
          new basic_irritator("irritator_egress " + id,
                if->uart_egress_generator,
                if->uart_protocol_bfm,
                if->uart_egress_checker);

      do_(if->uart_configuration->name);
      do_generator(if->uart_egress_generator->name);
      do_generator(if->uart_ingress_generator->name);

      count++;
    }
  }
}
```

Okay, that code is a bit long—but it is straightforward. First, a UART interface is chosen to be the focus of the test. Then it is built and randomized. After that, the rest of the UART interfaces are packed into test irritators and randomized.

Now that all the components have been built, let's look at a typical test method:

```
void quad_uart_irritators::start() {
  uart_test_component_ingress_->start();
  uart_test_component_egress_->start();

  for (teal::uint32 i(0); i < irritator_count; ++i) {
    uart_irritator_ingress_[i]->start();
    uart_irritator_egress_[i]->start();
  }
}
```

All the methods of the test follow this form. They first perform the action on the `test_component`, and then on the irritators. The `wait_for_completion()` is just a little different:

```
void quad_uart_irritators::wait_for_completion() {
  uart_test_component_ingress_->wait_for_completion();
  uart_test_component_egress_->wait_for_completion();

  for (teal::uint32 i(0); i < irritator_count; ++i) {
    uart_irritator_ingress_[i]->stop_generation();
    uart_irritator_egress_[i]->stop_generation();
  }
  for (teal::uint32 i(0); i < irritator_count; ++i) {
    uart_irritator_ingress_[i]->wait_for_completion();
    uart_irritator_egress_[i]->wait_for_completion();
  }
}
```

Notice that `wait_for_completion()` first waits for the focus of the test to complete. Then it tells the irritators to stop, and then waits for the irritators to complete.

Remember, that after this `wait_for_completion()` returns, the test is done.

The quad_uart_vectors Test

The `quad_uart_vectors` test is the logical evolution of the previous test. We get more "C++"- like. Instead of a fixed array, we use the template library's vector template. This is an appropriate use of the template library. Here are the relevant parts of the test header file:

```
class quad_uart_vectors : public truss::test_base {
public:
  quad_uart_vectors(testbench* tb, truss::watchdog* w,
                    const std::string& name);
  virtual ~quad_uart_vectors() {}

  //standard Truss methods not shown

 private:
  testbench* testbench_;
  uart::basic_test_component* uart_ingress_;
  uart::basic_test_component* uart_egress_;

  std::deque<truss::irritator*> irritators_;
  quad_uart_vectors (const quad_uart_vectors&);
  quad_uart_vectors& operator=
                         (const quad_uart_vectors&);
};
```

The randomize method is very similar to that in the previous test, with a small difference:

```
irritators_.push_back(
   new basic_irritator("uart_irritator_ingress " + id,
                          if->uart_ingress_generator,
                          if->uart_program_sfm,
                          if->uart_ingress_checker));
```

The methods all follow a standard form, but for those not familiar with the template library, they can look unnatural. Here is one example method:

```
void quad_uart_vectors::time_zero_setup() {
  uart_test_component_egress_->time_zero_setup();

  std::for_each(irritators_.begin(), irritators_.end(),
    std::mem_fun (&truss::irritator::time_zero_setup));
}
```

The template library `algorithm` file has a number of useful algorithms. In this case, this algorithm calls the method on all elements in the container.

The methods are a little more complicated for the Truss methods that have a parameter:

```
void quad_uart_vectors::report(const string p) const {
    uart_test_component_ingress_->report(p);
    uart_test_component_egress_->report(p);

    std::for_each(irritators_.begin(), irritators_.end(),
        std::bind2nd (
            std::mem_fun (&truss::irritator::report),
            p));
}
```

The `std::bind2nd` is a template class to make a multi-argument method look like a no argument method. While strange, this is a well-trodden path for C++ programming.

This last form of the test contains the fewest lines and uses standard C++ idioms. It's up to you as to whether this is appropriate for your system.

The uart_test_0 Test

The `uart_test_0` test is just a rework of the block-level test. The only changes were to use the testbench's `uart::interface` objects, and to select an interface to exercise.

Summary

This chapter took a look at a system-level verification system. We adapted the components from the block-level test.

The first test that was talked, just re-used the test_component from the block-level test on all four UART interfaces.

The next test brought in the concept of irritators, background traffic for a main test focus.

The system-level test `quad_uart_vectors` was used to show how the template library can be harnessed to make small, efficient standard-form code.

In general, this chapter showed that many of the block-level components could be adapted without modification to the code. We did, however, need to modify the `uart_16550_sfm` to handle a specific address range.

Things to Remember

"There goes my tail again." —Eeyore
Paraphrased from Winnie-the-Pooh, by A.A. Milne

An ending is, by definition, a new beginning. This, the last chapter, provides a good opportunity to review some of the handbook's main points. The authors sincerely hope that this is also a beginning for you to benefit from using some of the techniques presented in the preceding chapters.

This chapter is the 30,000-foot view of what we have covered. A wise, experienced manager once told the authors, "If you want your team to remember something, tell them at most three things." We take that advice—sort of—and present the three most important ideas of each part in the book.

We hope that this handbook, and its accompanying code, was and will continue to be useful. In the end, however, it is your job to verify the chip.

Part I: Use C++ and Layers!

In the first part of the book we introduced verification, C++, object oriented programming, and what a layered verification looked like. Here are the important points:

- C++ is a good language for verification.

- Use OOP techniques for verification, but not to excess.

- Layering is the main technique for a verification system.

C++ is what the majority of the software industry uses. As a result there are lots of books on C++, tools to support a C++-based development flow, and lots of open-source code. C++ has a rich and well-polished feature set. To learn C++, start slowly and add language features carefully.

The verification world is a bit enamored with OOP. We are probably in the early stages of settling down and using it, or not, where appropriate. By using OOP techniques we can communicate our architectural intent clearly.

The concept of layering, formally described as abstraction, roles, and responsibilities, is perhaps the single most important technique we can use. We presented terms for layers that we later implemented as classes and conventions.

Part II: An Open-Source Approach

In this part of the handbook we presented some code that has proved useful to us a nd those at other companies. That code may not have everything you want, but it should be flexible enough for you to adapt it to your needs. We noted specifically the following:

- Teal is a minimal, yet sufficient, interface to the HDL. It is the enabler when one uses C++ for functional verification.

- Truss provides a flexible, yet well-defined, verification application framework.

- A simple, but complete, example can be useful.

This part of the handbook is what most books lack. The authors take all the theory and lessons learned and show you how they have built verification systems. Make no mistake, Truss is a verification methodology. Teal is a bit more open, but any implementation of a programming concept contains the prejudices and biases of the implementors.

The point of the example is to show how these implementations, Teal and Truss, can be useful.

Part III: OOP—Best Practices

In this part of the handbook we took a long look at OOP. We talked about how to "think OOP" and how to "code OOP." Here are the three main points of this section:

- OOP is a powerful tool to manage complexity and create adaptable code.

- There are lots of techniques, and most of them involve balancing trade-offs.

- The code should make minimal assumptions, and make those assumptions as obvious as possible.

As the complexity of the chips increased, so did our verification systems. OOP can be used to increase the communication among engineers. Basically, this means creating code that others can reason about.

The hundred pages or so of the middle part presented lots of lessons learned. There were techniques, guidelines, and horror stories. There were no absolute right or wrong answers. You and your team must decide what is appropriate.

If a bit of code has a well-defined purpose as well as obvious dependencies, it stands a good chance of being reasoned about and eventually understood. The objective is to minimize the assumptions about the code, while still doing something worthwhile.

Part IV: Examples—Copy and Adapt!

We could have left the book with only three parts—but one example is usually not enough to help you understand a set of techniques or some new code. Consequently, we wrote some more examples. The following summarizes the main points of these examples:

- You can create portable verification IP that other projects can use.

- Separating the chip-specific parts from the protocol-generic parts shows users what they have to modify for their project.

- The testbench and test components can become large, but they are still "reasonable."

This section of the handbook presented more examples of chips and their verification systems, all the way to a final example that used all the previous examples. We would not be surprised if the code has mistakes and can be made even more clear. Yes, we'll probably even get some complaints, but we know of no better way for you to learn the ideas and techniques talked about in this handbook than to see working, completed examples.

Conclusion to the Conclusion

The authors have tried to make a handbook that is useful. We've combined verification and C++ and described techniques that have proved useful.

We did not separate the verification techniques from the language used to express them. To do that would have made the book easier to write. However, you would have been reading just a book, not a handbook. You should be able to find in these pages—and on the accompanying CD—enough examples that are sufficiently close to what you want to do. Cut, copy, and paste away!

If you want to contact us, we are at www.trusster.com. On this site you can also find up-to-the-minute information about Teal and Truss, as well as discussion boards where users share knowledge and ideas. It's a good place to start for any Teal or Truss questions.

It's also where we will post errors found in this handbook. We invite your comments and suggestions.

". . .and now for something completely different.[1]"

[1.] From Monte Python's Flying Circus, episode 26, December 1971.

Index

Symbols

base class 45, 178, 205
basic_test_component 312
baud rate 253, 284
baud_rate 283
beauty 235
befuddlement 155
Benjamin Franklin 193
BFM 16, 36, 50, 96, 177, 195, 205
BFM agent 205, 222
bfm_agent 206
bidirectional protocol 292
bin 123
binary operations 34, 210
bins 14
bit fields 74
Bjarne Stroustrup 1, 22, 41, 191
black-box function 69
block length 301
blocking method 227
block-level testbench 308
BOOK_HOME 279
boolean 111, 255
    operations 74
boot source 169
boot_source 169
bottom-up approach 47
brittle code 219
broadcast mechanisms 52
build() 262
built-in types 34, 38, 81, 196, 210, 211
burst 201
bus 245
    contention 105
    functional model 16, 36, 50, 96, 177, 205

C

C 21, 22, 27, 288
C with Classes 22, 156
C++ creator's home page 42
C++ Primer 42
C++-to-HDL gasket 51
C++-to-HDL interface 5
cache coherency unit 185
cached 262
cached data 242
Cadence "e" 18
Cadence Specman "e" 12
callbacks 230
CAN 241

CAN node 243
can_fifo 243
can_node 241, 262
canonical form 212
capitalization 260
case 180
catch handler 215
c-function 86
change 86
channel 128, 211, 220, 231
    class 111
    connection 231
    policy 293
channel::get_data() 220
channel_get 128
channel_put 128, 296
channels 291, 296
char 38, 80
chars 15
check loop 149
checker 53, 106, 185, 187, 195, 220, 224, 267, 279,
    294
    agent 292
checker.cpp 148
checker_agent 294
checking 147, 187, 267
checking side 293
chip programming 286
class 23, 194
    abstract base 24
    base 45
    container 38
    ethernet_monitor 49
    factory method 262
    inheritance 34, 171
    library 71
    monitor 48
    names 257
    pci_express_monitor 48
    reg 71, 74
    testbench 54
    verification_base 37
    virtual base 24
    vlog 71
    vout 71
    vrandom 72
    vreg 71, 74
classes 29, 43, 196
class-referencing mechanism 197

L

M